Introduction to the Engineering Profession

Second Edition

Introduction to the Engineering Profession

Second Edition

M. DAVID BURGHARDT
Hofstra University

HarperCollins*CollegePublishers*

To Linda

Sponsoring Editor: John Lenchek
Project Editor: Cathy Wacaser
Design Administrator: Jess Schaal
Text Design: Lesiak/Crampton Design Inc: Cindy Crampton
Cover Design: R.R. Gonzales
Cover Photo: *Background Image:* © Patrick Doherty/The Image Bank
   Inset Image: © Comstock
Picture Researcher: Corrine Johns
Production Administrator: Randee Wire
Compositor: Interactive Composition Corporation
Text Printer and Binder: R.R. Donnelley & Sons Company
Cover Printer: R.R. Donnelley & Sons Company

Introduction to the Engineering Profession, Second Edition
Copyright © 1995 by HarperCollins College Publishers

All rights reserved. Printed in the United States of America. No part of this book may be used or reproduced in any manner whatsoever without written permission, except in the case of brief quotations embodied in critical articles and reviews. For information address HarperCollins College Publishers, 10 East 53rd Street, New York, NY 10022.

Library of Congress Cataloging-in-Publication Data

Burghardt, M. David.
   Introduction to the engineering profession / M. David Burghardt.—2nd ed.
      p.   cm.
   Includes index.
   ISBN 0-673-99371-X
   1. Engineering—Vocational guidance.   I. Title.
TA157.B87 1994
620′.0023′73—dc20   94-26512
      CIP

96 97 9 8 7 6 5

Contents

Preface   xi

CHAPTER 1   **SUCCEEDING IN ENGINEERING**   1
- 1.1   Introduction   2
- 1.2   Creativity in Engineering   3
- 1.3   Technology and Its Political Implications   4
- 1.4   Why Do I Have to Study This?   6
- 1.5   Your Curriculum   10
- 1.6   Minimizing Your Study Time   12

References   16
Problems   17

CHAPTER 2   **HISTORICAL PERSPECTIVE**   20
- 2.1   Introduction   22
- 2.2   Patterns of Change   22
- 2.3   Beginnings of Engineering   25
  - Egypt and Mesopotamia   25
  - Greece and Rome   29
  - Dark Ages and Middle Ages   33
- 2.4   The Developing Industrial Age   34
- 2.5   Engineering Societies in the United States   39

References   41
Problems   41

CHAPTER 3   **FIELDS OF ENGINEERING**   44
- 3.1   The Technological Team   46
- 3.2   Engineering Societies   48
- 3.3   Accreditation Board for Engineering and Technology (ABET)   51
- 3.4   Fields of Engineering   51
  - Aeronautical and Aerospace Engineering   52
  - Agricultural Engineering   53

　　　　　Architectural Engineering　　54
　　　　　Automotive Engineering　　54
　　　　　Biomedical Engineering　　55
　　　　　Ceramic Engineering　　55
　　　　　Chemical Engineering　　56
　　　　　Civil Engineering　　58
　　　　　Computer Engineering and Computer Science　　59
　　　　　Electrical and Electronics Engineering　　60
　　　　　Environmental Engineering　　62
　　　　　Industrial Engineering　　62
　　　　　Manufacturing Engineering　　63
　　　　　Marine Engineering, Naval Architecture, and Ocean Engineering　　64
　　　　　Materials and Metallurgical Engineering　　65
　　　　　Mechanical Engineering　　66
　　　　　Mining and Geological Engineering　　67
　　　　　Nuclear Engineering　　67
　　　　　Petroleum Engineering　　68

3.5　Technical Careers in Engineering　　68
　　　　　Research　　69
　　　　　Development　　70
　　　　　Design　　71
　　　　　Manufacturing and Construction　　72
　　　　　Operations and Maintenance　　73
　　　　　Sales　　74
　　　　　Management　　74
　　　　　Government, Consulting, and Teaching　　75
　　　　　Quality Assurance and Quality Control　　76

3.6　What Is Quality?　　76

References　　85
Problems　　86

CHAPTER 4　ETHICS AND PROFESSIONAL RESPONSIBILITY　　88

　　4.1　Introduction　　90
　　4.2　What Is a Profession?　　90
　　　　　Complex Subject Matter　　90
　　　　　Certification of Competence　　91
　　　　　Trustworthiness　　91
　　　　　Professional Organizations　　92
　　4.3　Moral Dilemmas　　92

4.4 Product Safety and Product Quality — 94
Ethical Problems 96
The DC-10 Disaster 96
Nuclear Reactor Welds 97
Being Right Is Expensive 98
A Positive Side to Perseverance 99
Holding the Line 100

4.5 Design Changes — 101
4.6 State-of-the-Art Is Not Always Enough — 102
4.7 Safety and Risk—Risk Assessment — 102
4.8 Situations You Will Face — 106
4.9 Case Studies — 107

References 110
Problems 111

CHAPTER 5 COMMUNICATION—WRITTEN AND ORAL 114

5.1 Introduction — 116
5.2 Function of Communication — 116
5.3 Written Communication — 117
5.4 The Written Report — 118
5.5 Résumé or Vita — 121
5.6 Memos — 124
5.7 Oral Communication — 124
5.8 Library Usage — 127

References 131
Problems 131

CHAPTER 6 GRAPHICAL COMMUNICATION 134

6.1 Introduction — 136
6.2 Presentation of Qualitative Results — 137
6.3 Presentation of Quantitative Results — 142
Line Graphs 142
Equation of a Straight Line 146

6.4 Plotting on Semilog and Log-Log Paper — 148
6.5 Sketching — 154

References 157
Problems 157

CHAPTER 7 STATISTICS AND ERROR ANALYSIS 164

- 7.1 Introduction 166
- 7.2 Measures of Central Tendency 167
- 7.3 Measures of Dispersion 169
- 7.4 Probability and Normal Distribution 170
- 7.5 Linear Regression Analysis 177
- 7.6 Error Analysis and Error Propagation 179

References 182
Problems 182

CHAPTER 8 CONCEPTS OF PROBLEM SOLVING 186

- 8.1 Problem Format 188
- 8.2 Problem Solving 190
- 8.3 Significant Figures and Scientific Notation 193
- 8.4 Unit Systems 195
  - Fundamental and Derived Units 195
  - English Engineering Unit System 197
  - SI Units 199
- 8.5 Conversion Factors 203
- 8.6 Technical Decision Analysis 204
  - Decision Tree 205
  - Multicriteria Decision Analysis 206
  - Gantt Chart 207

Reference 208
Problems 208

CHAPTER 9 ENGINEERING DESIGN 212

- 9.1 The Design Process 214
- 9.2 Problem Definition 215
- 9.3 Creating Solutions 216
- 9.4 Refinement and Analysis 219
- 9.5 Problem Resolution and Implementation 221
- 9.6 Additional Considerations in the Design Process 224

References 227
Problems 227

APPENDIX 1 **NSPE CODE OF ETHICS FOR ENGINEERS** **233**
  Preamble   233
  Fundamental Canons   233
  Rules of Practice   234
  Professional Obligations   236
  As Revised March 1985   239
  Statement by NSPE Executive Committee   240

APPENDIX 2 **CONVERSION FACTORS**   **241**

APPENDIX 3 **COMPUTERS AND COMPUTER APPLICATIONS**   **250**
  A3.1 The Computer Processing Cycle   **250**
  A3.2 Computer Applications   **255**
    Word Processing   257
    Spreadsheets   257
    Computer-Aided Design   258
    Data Bases and Data-Base Processing   263
  A3.3 Computer Programming Languages   **264**

APPENDIX 4 **ALGEBRAIC AND TRIGONOMETRIC PROBLEM SOLVING**   **268**
  Linear Equations   269
  Simultaneous Linear Equations   272
  Linear Systems with Three Variables   274
  Matrix Solution of Linear Systems   277
  Quadratic Equations with One Unknown   280
  Exponential and Logarithmic Functions   282
  Trigonometry   284
  The Laws of Cosines and Sines   287
  Circular Sine and Cosine Functions   289

  Index   293

Preface

Introduction to the Engineering Profession is designed to help beginning engineering students decide on their field of engineering, gain a perception of the history of engineering and their place in it, and develop the survival skills necessary for an education and career in engineering. An underlying theme of the text is that society needs people who are educated as engineers to assist in the governance of this technologically complex social system, which in turn requires that engineers become more socially and politically aware and active.

Many students use an introduction to engineering course to better define a field of study about which they are uncertain. They are seeking reasons for studying engineering. A number have preconceived ideas about the profession, though few understand the breadth available to them. Of those students with prior knowledge, some truly want to be design engineers, and most want to earn a good income and eventually to move into management. This latter issue is addressed early on in the text.

Introductions to engineering courses are highly individualized, varying significantly from school to school. Therefore, this text provides the key elements for a variety of introductory courses.

The character of engineering education is changing, reflecting a greater professional concern about the social and political consequences of technological change. Certainly this message must be transmitted to future generations of engineers, and it is one of the philosophical underpinnings of the text, reinforced in virtually every chapter. Not only is the character of the profession changing, but the people entering the profession are different from those who entered it in the past. The stereotypical white male engineer is being joined by people from all races and both genders. The engineering workplace will benefit from the infusion of people with different backgrounds, and our educational programs must address new issues necessary to incorporate these students into the engineering profession.

In the past, engineering educators often presumed students had a personal familiarity with the engineering profession, either through family members or friends. The students entering engineering education today often have no immediate role models that define what an engineer is. In this text we attempt to provide such role models.

The underlying purpose of Chapter 1 is to give students techniques for succeeding in college and realizing their potential. Frequently, students who leave the engineering field do so because of inadequate study habits rather than lack of aptitude or desire. Acquiring good study habits at the outset of their education could help them succeed in engineering. Incorporated throughout the text are comments from practicing engineers regarding the profession, on topics ranging from work habits to communicative ability to the joy of designing a new product.

To understand the engineering profession today, we need to have a historical view of its development. Not only does the material in Chapter 2 trace the evolution of civilized society, but it also describes the evolution of the engineering profession as seen through the formation of engineering professional societies in the United States in the late 1800s and early 1900s. The dual allegiance to society as well as an employer caused much conflict in those times and still creates moral dilemmas for engineers today. Furthermore, in the early 1900s engineers were expected to provide solutions to society's problems, and this theme is reemerging today. Perhaps recognizing past pitfalls will help us avoid similar ones in the future

No introductory course in engineering would be complete without a section on ethics and professional responsibility. Chapter 4 presents problems and guidelines that can serve as a springboard for class discussion. An expanded discussion of risk assessment including factors perceived as causing risk also is included in Chapter 4.

Some engineering majors decline to think of communication (represented by their humanities and social science courses) as being essential to their success as engineers. Chapter 5 attempts to debunk this view and also supports recent publications of the National Society of Professional Engineers (NSPE) regarding the need for engineers to be able to communicate well. Because English composition courses often do not address technical report writing, this topic is covered in Chapter 5 to give students the background necessary to write the many reports required in their undergraduate careers. Following graduation, engineers often communicate through memorandums, so a discussion of memo writing is included, with emphasis on the design memo.

At first glance it may seem that the topic of résumé preparation is out of place in an introductory course; however, an engineer's career begins with this course, not after completion of the last course during senior year. The résumé assists in career planning starting in the freshman year. Chapter 5 also covers technical library usage, another valuable asset to students.

An essential element to virtually all introductory engineering courses is a discussion of the importance of engineering design. The various aspects of the design process are detailed in the new Chapter 9, Engineering Design. This chapter examines some techniques and procedures that help in creating new designs and includes a section on design for the environment. Homework problems include use of the design memo.

Concurrent engineering is discussed in the text with examples drawn from industry illustrating its advantages as well as the problems involved in getting management and workers to adopt this system.

A new appendix, Algebraic and Trigonometric Problem Solving, has been added as well. Many engineering students have difficulties with algebraic word problems, not with engineering concepts. The numerous example problems in the appendix as well as others in the text use a stylized engineering problem solving format. The hope is that by formulating the problem in their own words, drawing sketches of the problem, and stating assumptions, students will improve their algebraic and engineering problem solving abilities.

The overall focus of the text is to interest and excite students about the various opportunities afforded by an engineering education: the creative challenge and reward of engineering design, the use of analytic skills in problem formulation and solution in engineering and management, and the skills and attitudes necessary for a rewarding and stimulating college education and for a satisfying career after graduation.

Introduction to the Engineering Profession

Second Edition

CHAPTER 1

Succeeding in Engineering

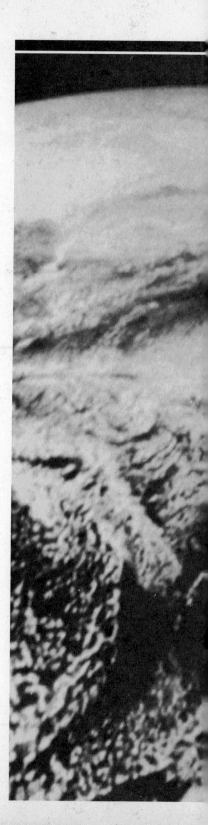

CHAPTER OBJECTIVES

To learn what engineering is and is not.

To identify creativity as an essential element of an engineering education.

To distinguish between the concepts of technology and engineering.

To understand why your course of study in engineering is structured as it is and contains the courses it does.

To make use of techniques that will minimize and improve your study time.

(Photo courtesy of NASA.)

You are on the threshold of a very exciting career in engineering, at a time when the world needs the best technologically aware and creative people possible. People who can solve the multitude of environmental problems that confront us and who can bring us closer together in a world community will lead our future.

1.1 INTRODUCTION

What is engineering? How do you succeed at it? You are embarking on a potential career in engineering, and with the help of this course and text and all the other courses and texts you will have, a definition of engineering will evolve. It may be one of the most useful majors a student can select among all the choices available at most colleges and universities. It is a course of study followed by a professional life devoted to the creative solution of problems. To succeed in any endeavor you must fully utilize your natural abilities, and in the end you will judge your own success on how well you accomplished this. The world that envelops us is a technological one. Think about all the technology you encountered before you left home this morning.

The alarm clock wakes you. Not only is the clock run by electricity, which must be produced somehow, somewhere, but someone, somewhere, designed the alarm mechanism.

You take a shower. The faucets are valves, which were designed by an engineer. The water comes through piping from a central water supply. The entire water distribution system was conceived and designed by engineers.

Opening the refrigerator for milk and juice, you realize that the refrigerator had to have been designed and built according to engineering specifications.

You can expand this list and should do so to gain a perspective of how totally our lives are surrounded by and dependent on technology. No longer do we rise with the sun and depend on manual labor in the fields or forest for our livelihoods; a more sophisticated existence is required. If one does not understand the technology that governs so much of our support systems, a sense of unease or helplessness develops. An underlying advantage of your engineering education is that you will understand the myriad technologies that govern so much of our lives, and thus not be controlled by them.

We all have a limited understanding of what an engineer is. Through education and with the practice of engineering our perception of engineering and what it means to be an engineer will expand. The term *engineer* conjures up many images. Who is your favorite engineer? An aunt or uncle, a parent, cousin, or friend? A famous inventor from history? An unusual engineer whose works are enjoyed by many is Alfred Hitchcock. Although he is best known as a movie director, he was trained as a mechanical engineer. Contrast the crisp sequencing in a movie like *The 39 Steps* or *North by Northwest* with the more ephemeral movie, *A Man and a Woman*, directed by Claude Lelouch. The influence of engineering is apparent. We will better understand the close link that engineering has to other creative fields later in this chapter.

One of the difficulties we often face when uncertain about selecting a course of action, such as studying engineering, is that others can dissuade us from that course with thoughtless remarks. It is demoralizing to suffer comments such as "Why do you want to study that?" or "That's too hard," or even "Engineers are nerds." As we will learn more fully in Chapter 3, part of the reason for this perception is that engineers often do not remain practicing engineers, and hence people do not directly see the value of an engineering education. Would anyone question your desire to follow in the footsteps of engineers such as Thomas Jefferson, Benjamin Franklin, David Sarnoff, Neil Armstrong, Lee Iacocca, or Jimmy Carter? Society needs people educated as engineers to help solve the problems that daily confront our lives. Historically engineers have been trained not only to solve technical problems but also to have the techniques and the understanding at their disposal to solve nontechnical problems as well. This will be part of your challenge in life. This will provide your response to provocative comments, your reason why the hard work of an engineering education is well worth the effort.

1.2 CREATIVITY IN ENGINEERING

One of the reasons a career in engineering can be so rewarding is that you are compensated very well for creating. Creating is one of the fundamental acts of our existence that gives us great pleasure. Observe small children, or reflect back on your own early childhood, and see the excitement and the development of self-esteem that creating provides. Children are constantly making things—necklaces, castles, cars, rockets—in endless variety. As we mature we become more self-aware and self-critical, fearing to create because others will think it not worthy. The effects of peer pressure in our teens can be very detrimental to creative development. The desire to create remains, however. Your engineering education encourages you to be creative in designing a new product, be it a building, a machine, an electronic circuit, or computer software. With the techniques and knowledge gained in your course of engineering study and with your innate creative ability, something that never existed before can be designed—created. Thus, the general purpose of your engineering education will be to provide you with technical tools and to encourage the use of your own creativity, culminating in a new design that will solve a given problem.

Engineering is strongly associated with other creative fields, particularly the fine arts. Chapter 2 analyzes engineering from a historical perspective, but here let's leap back in time for just a moment to more primitive eras, before recorded history, when engineering existed in a very fundamental form. We must look to the beginnings of civilization and the forms of creativity exhibited.

Art or engineering? In actuality this is a water tower used in a municipality's water supply system. (Courtesy of Sidney B. Bowne and Son)

How did people create? What is the definition of art? Art is simply something we create, the results of our creativity. Fine art, often confused with the more general term *art*, is a manifestation of creativity with no functional purpose, only aesthetic purpose. Crafts are manifestations of creativity with both functional and aesthetic purpose. The originators of engineering were craftspeople, such as flint knappers, whose understanding of different types of material, such as flint, and its intrinsic properties allowed them to refine nature and create weapons and tools. In the fine art of painting, artists must learn the technical skills of color usage, different application techniques, painting on different surface textures, and the ability to see what exists, before connecting with their innate creative powers and producing what is called a work of art.

Engineers face a similar problem, in that the techniques are the analytic tools with which we must be comfortable, before connecting with our creative powers and producing what is called a new design. Just as the most difficult part of an art student's education is to learn to see what is, rather than their precognition of what is, the most difficult part of an engineering student's education is to learn and feel comfortable with the analytic tools of engineering. When this is accomplished, connecting with your creative being readily occurs.

1.3 TECHNOLOGY AND ITS POLITICAL IMPLICATIONS

The term *technology* is often confused with *engineering*, probably because engineers create it. Technology is the manifestation of engineering creativity; it results from creativity with a purpose, or engineering design. Consider the problem of sending someone to the moon and returning to earth in 1968. The technology to do so did not exist; it had to be created. Created by engineers.

Rocket shielding materials that did not exist had to be created to withstand the tremendous temperature generated by reentry into the earth's atmosphere. Computers were created to control the rocket's engines, which also were new. The list is very long. The point is that new technologies are being created all the time by engineers solving a given problem. These technologies may be adapted by other engineers to solve a different problem. Thus, the material advances in ceramics used for shielding rockets are now being applied in internal combustion engines.

Engineers are often involved in solving problems that have political aspects. Consider the following situation. The town you live in has been using a landfill as a way to dispose of its garbage. The garbage is dumped and then buried with sand and dirt. The capacity of the landfill area will soon be exceeded, since the garbage can only be piled so high. Also, the chemicals from the garbage are leaching into the groundwater, and hence this method of disposal can no longer be used.

Two alternative methods are available, high-temperature combustion of the garbage, or recycling many of the garbage components and using landfill or incineration of the nonrecycled remainder. The technology for both exists. The heat from the incineration of garbage is used to produce steam, which in turn drives turbines and generators, creating electric power. In this situation the electricity generated partially pays for the incineration process. A major problem is that the incineration of some plastics causes the formation of dioxins, which weaken the immune system of the human body. Metals, such as aluminium, and paper and plastics could be recycled as well. If recycling were combined with incineration, the remaining burnt waste could be buried or used as landfill. The removal of paper and plastic reduces the garbage's heating value, so not as much electricity could be produced, and the cost of

A waste recovery plant incinerates solid waste and generates electricity. This plant has a burning capacity of 2300 tons per day of waste, producing 72 megawatts of electricity. (Courtesy of American REF-FUEL)

the process and your taxes rise. There is also an additional cost in removing the recycled material and selling it, and often little or no market exists for recycled material. Other difficulties are having the populace segregate the garbage so it can be recycled, the cost of developing a recycling machine, and finding a market for the recycled material. In 1985 in New York State, two-thirds of the plastic bottles collected under the bottle deposit system were buried, rather than recycled, because the market conditions made this the most cost-effective alternative.

What exists is a sociological problem needing a technical solution. Research is underway to find a method of combustion of garbage, with plastics included, that will prevent the formation of dioxins, or remove them if they are formed before they are released to the atmosphere. One option is to choose incineration, and hope that a technical cure will solve the dioxin problem, or accept the limited amount of it as an acceptable risk; a second is to set up a recycling system that requires consumers to separate materials at the pick-up point, or install machinery to do this at a central depot, and have a market for the recycled material. You certainly would not want to stockpile the material. Even in the recycling and incineration system there will be some plastic burned and some dioxin formed, but significantly less than in the straight incineration system.

Both systems require engineering analysis, not high tech engineering necessarily, but traffic flow, storage of materials, plant location, incinerator design. While your mind is focused on this garbage disposal problem, think about the political and technical problems associated with plant location. There is a term that politicians use when everyone wants something, but no one wants it near them—NIMBY, not in my backyard. Engineers must become more attuned to the political aspects of technical solutions.

1.4 WHY DO I HAVE TO STUDY THIS?

All engineering curricula are quite similar in that the latitude available to you in course selection is rather circumscribed compared to nonengineering majors, such as history. Part of this stems from your education being counted as professional experience when it is time for you to be certified as a professional engineer (covered in more detail in Chapter 3) and part stems from the fact that learning the material covered during your senior year requires having taken a sequence of course in previous years. Engineering requires a vertically integrated education, where more advanced courses depend on information learned in preceding courses. This is contrasted with horizontally integrated curricula, where learning the information in one course does not require the prerequisite of another course. For instance, your understanding of the history of modern China does not depend on your understanding of that of Latin America, though, of course, one may enhance the other.

Let's consider the following situation as an example in explaining certain aspects of a mechanical or civil engineering curriculum. A deep stream that is five feet wide needs a bridge so people can walk across it. You find a plank two inches thick, twelve inches wide, and eight feet long and place it on the two banks, crossing the stream (Figure 1.1). No engineering education was

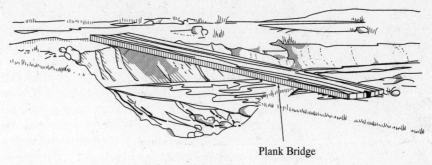

Plank Bridge

Figure 1.1 A plank bridge across a stream.

needed in this case; from your experience you decided this should do the job. In fact, you are so pleased that you decide to go into business making plank bridges. But now people who might buy the bridges have questions about how great a load (weight) the bridge can support. Will it support six people? What size people? you ask. And so it goes. From an engineering viewpoint, what exists in this case is a beam supported on two ends, and in this simplest view a rigid beam that will not deflect.

You need to analyze this rigid rigid beam. To do this you must model the beam with mathematical equations, which derive from the calculus and differential equations you will be studying. You also need to understand the physical situation, hence the courses in physics you are taking. Then you must take a course in engineering mechanics, where you learn to predict the various forces throughout the beam that are caused by the weight of the people. Finally, for the simplest case, after a course in strength of materials you can determine whether or not a wooden plank with certain assumed properties will break under the aforementioned forces.

Laboratory experiments often create situations where you can work directly with faculty members, an ideal time for one-on-one learning. (Courtesy of Hofstra University)

Remember at this point you are analyzing what is assumed to be. You are not analyzing the wooden plank that exists: you have developed a mathematical model of that plank that presumes that the wood has certain properties, and from this you determine whether or not the plank will sustain the load. You feel certain that it can hold 750 pounds, but to be on the safe side you guarantee it will support 500 pounds. What you have inadvertently introduced is a "factor of safety," which all designs have and which accounts for unexpected load conditions or property variations. We are using the word *design* a little loosely in this case, as you did not really design the plank but assumed it to be sufficient. Another aspect of this situation is that although the model will support 750 pounds, the actual plank might not. Your analysis assumed that the wood had certain properties and could be modeled in a certain mathematical fashion. Perhaps a knot located in the center of the plank causes it to break with a load of 500 pounds. With advanced engineering courses, having as prerequisites additional mathematics and engineering courses, you will be able to account for knotholes. An equally important aspect is that what is designed and analysed as being satisfactory should be what is built. Very often the disasters we read about are caused by a difference between the design and the actual construction. If differences do occur in construction or in materials used, the analysis should be performed again to assure that the de facto new design, what was actually constructed, meets the requirements of the work specifications.

Now a competitor enters the picture, as someone else decided that you are not the only one who can lay planks across a stream. To be competitive and create a better product, you wish to design something new. There could have been cost-saving tactics used first, such as shortening the plank, so there would be less support on each bank, or reducing its thickness, or even reducing the factor of safety, but let's not deal with these at this point. Using your knowledge of strength of materials and engineering mechanics, you create a new bridge design; it is modeled and analyzed to see if it meets the load requirements and the factor of safety you wish. Thus, there is a significant difference between design and analysis; both are important aspects to your engineering education but one without the other is not sufficient in the education of an engineer.

Let's go back to the competitive situation that now exists between your business and others in the same field. Being an ethical person, you selected a very conservative factor of safety. Your competitor chose a smaller factor of safety and feels that it is quite adequate; that means his or her bridge, if of a similar design, will cost less and hence be selected. Municipalities and states have recognized this competitive problem and have established building codes, which define the minimum requirements for a variety of construction situations. Very often the contract specifications will detail the requirements that the design must meet in addition to statutorial ones. This means that everyone is operating from the same vantage point. The problem with this is that the codes and statutory requirements are based on existing designs, and hence design innovation can be restricted by these codes. This is particularly true when new materials allow designs that were not previously possible. Again, technical solutions merge with political realities as special interest

1.4 WHY DO I HAVE TO STUDY THIS?

The construction of the Hoover Dam on the Colorado River was an immense undertaking in civil engineering. During the years it took to pour the concrete, the river had to be temporarily diverted. The dam creates a recreational lake upstream, Lake Mead, and generates electricity. (Courtesy of ASME)

groups develop to try to keep the laws the same so they can continue their existence.

As we continue with this story, let's imagine that your designs are creative and provide you with a product that is better than your competitor's. Business expands, and you hire people to work for you—other engineers and people to fabricate the bridges that you design. Are you a business executive or an engineer? Where is your identity?

Let's change the scenario, using an electrical circuit design to illustrate some other concepts. You are an electrical engineer and work for a small business that manufactures Christmas tree lights. In its thirty years the company has not changed the design from that shown in Figure 1.2. This is known electrically as a series circuit, where the lights are connected in a row, or series. When the string of 20 lights is plugged into the wall, a voltage potential, like a pressure, causes electrical current to flow through the lights and back to the plug. This is the simplest design, but it contains some flaws. If any of the bulbs are defective, then no current can flow through the circuit. An open circuit is created, as if the wire were served, and the entire string of lights will not work. It is also difficult to determine which of twenty bulbs is defective. A second design deficiency is that the brightness of the lights varies with the number of lights on the string. The more lights, the dimmer each glows. This happens in your home when you turn on a hair dryer in the bathroom and the lights dim because there is more resistance, load, on the circuit. You are aware of these limitations and propose a new design, as illustrated in Figure 1.3 (the bulbs are represented by sawtooth lines). In this

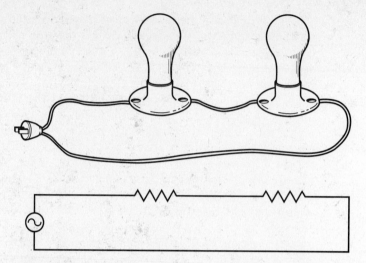

Figure 1.2 An electrical circuit with resistors (lights) connected in series.

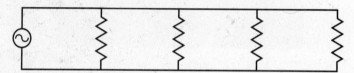

Figure 1.3 An electrical circuit with resistors (lights) connected in parallel.

case the lights are connected in parallel, and whenever a bulb is defective, the others remain lighted. Furthermore, the voltage across each is the same and the brightness is the same, regardless of how many bulbs are added to the string. This design is more reliable than the previous one. A failure of one of the circuit elements does not prevent the other elements from working. Reliability is an important design concept and moves us in the direction of product quality, a topic discussed in Chapter 4. (As your course of study in electrical engineering guides you through very complex circuits, you may want to think about designing a circuit that will cause the lights to blink on and off, randomly or periodically.)

1.5 YOUR CURRICULUM

Why am I studying these courses? Do I really need these courses? These are questions that you undoubtedly have asked or will ask as you pursue your education. Of course they cannot be answered specifically here, but in general the following will give an overview from which you can assess the curriculum of your major.

Your engineering curriculum provides you with the mathematical skills and understanding of the natural sciences that you will need in your later engineering course work. Usually your freshman year is devoted to learning calculus, chemistry, and physics as well as courses introducing the engineering profession, graphical representation, and computer programming. Additionally you start to fulfill requirements in English and social science and humanities; these courses are important for increasing your understanding and awareness of the sociological and political implications of technological change. The importance of good communication skills cannot be underestimated in the practice of engineering. As we will examine in Chapter 5, engineers must be able to communicate effectively in written reports and memos as well as orally. As much as one-third of an engineer's time is spent in reading and writing reports and memoranda. Investigate "want ads" in a newspaper and observe that good communication skills are mentioned in many.

The freshman year is very similar for most engineering majors, and it is often not necessary to decide on a specific major until the year is completed. Your curriculum has a flow: the mathematics is necessary for the physics and chemistry, the physics and chemistry for the engineering courses. More mathematics is required for the engineering courses, which is why your course of study is so mathematically intensive. You will model nature with the math, and as your models become more sophisticated, so must your mathematical understanding.

Two elements to your engineering courses are engineering science and design. Your courses in the sophomore and junior years will be primarily engineering science; you will be learning analysis techniques. These courses typically include material science, thermodynamics, circuit theory, and engineering mechanics. In your later courses you will learn more about analysis and design and will make decisions in the problem solution. An engineer must be able to choose among alternative solutions, based on a set of criteria, and your education tries to give you this experience in the later courses, such as machine design, thermal engineering, microprocessor systems, and structural design.

Additionally, you will have technical electives to tailor your course of study to the area of greatest interest to you. Perhaps as an electrical engineer you will specialize in communications, as a mechanical engineer in controls, or as a civil engineer in structures. Most often you will have a capstone design course or courses, in which you will design a product or process from start to finish. In aerospace the aircraft design course considers the optimized design of an aircraft meeting the specifications of payload, range, cruising speed, and runway length. The aircraft characteristics, such as wing shape, are determined and then analyzed and refined. Similar courses exist for other majors, and aspects of design are included in most senior-level engineering courses.

A practicing engineer reflecting on his career offers the following:

"An engineering career can encompass many aspects of the profession. Therefore, I would encourage all engineering students to get as broad based an education as available and apply themselves to all sbjects. You never know where you will be careerwise in five to ten years and subjects that you think will never be utilized can have direct applications."

Students often work in teams when performing laboratory experiments. (Courtesy of Hofstra University)

An engineer with 20 years' experience offers the following:

> "An engineer must be prepared for change. Today's engineer and most importantly tomorrow's engineer has to be adaptive and geared for change. This will require that the engineer stay abreast of technology shifts and changes. In order to accomplish this, the engineer will have to make a firm commitment to continuing education. It is important that the engineer realize that his/her education is never complete. There are many avenues available to accomplish this such as formal education, industry training workshops and professional seminars."

1.6 MINIMIZING YOUR STUDY TIME

Few people like to study; there is always something more exciting to do. For students, however, studying is part of the job, a necessity for being a well-educated person. If you have to study, and sometimes it is fun learning new ideas, then at least study as efficiently as possible, creating the opportunity for other activities in your life.

The study habits you had in high school will in general not be good enough for college. In high school the competition was different; not all students were college bound. Now that you are in college, the average ability of your peers is higher, so your expectations of yourself must improve. One of the differences between college and high school is the lack of outside forces, such as teachers and parents, encouraging your studying. Also, the assignments are less repetitive, so there is less chance to ignore certain material, lounge about, and expect to pick it up later. This twofold difference, no external encouragement and less repetition, is often why freshmen have a difficult time adjusting to college life. For engineering students this is particularly trying, as the material within a course is often vertically integrated, depending on the previous day's assignment, and without persistent attention it is very easy to fall drastically behind. When this happens, you feel discouraged and tend to want to change majors to something easier, though not necessarily better, for you. One of the requirements for surviving an engineering education, and a quality that many businesses look for in their employees, is that of persistence. There is a trite but true saying, that engineering is more perspiration than inspiration. Your course of study and career will require you to be diligent and persistent; you need not be brilliant to be a fine engineer.

One of the best mechanisms to help you allow enough time to study is creating a schedule for yourself, allotting time for a variety of activities. You cannot expect your entire college life to consist of eating, going to class and studying. An important aspect of your education is participation in extracurricular activities, which may or may not be associated with engineering. This is particularly important because engineers must take a more active role in social and political functions. We develop these abilities and interests through contact and practice in college. You might be interested in student government, for instance, and volunteer to work on committees. Perhaps the dormitory you live in needs help in organizing and running activities for its members. Join and participate. However, you need to schedule these activities and your work activities (study) as well. The system you are creating is complex, and you need to manage it well.

You could use an $8\frac{1}{2}'' \times 11''$ piece of paper with the hours of the day and the seven days of the week listed; Figure 1.4 shows an example. It may seem

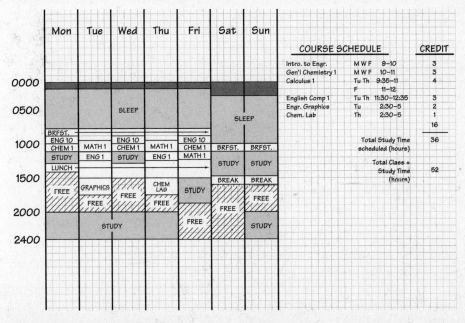

Figure 1.4 A chart of the hours in a week, separated into study time, class time, and other activities.

redundant to say the seven days of the week, but many students tend to think of school as a five-day-a-week job. It isn't at all. You must schedule the entire week, as weekends will frequently be spent in report writing and studying.

It is not as bad as you may think. Just think, there are 24 hours to the day, 168 hours per week. Assume that you spend 11 hours a day sleeping and eating, which leaves 91 hours for classes and studying. First, insert your class schedule. Then look at the time around the classes for other required activities—exercising, commuting, team practice if you are involved in sports. You should estimate that, for each class hour, you will be spending or should allot two to three homework hours. Lay this out on the schedule. Suddenly, there are gaps in the schedule where you are not doing anything. Often this occurs between classes; look to use this time effectively. "Effectively" may mean attending to student organization work, copying over notes, socializing with other students, but not focused styding. Whenever you can put together two or more consecutive hours, then you can have a focused study effort. Figure 1.4 indicates this. You do not have to stick to the schedule rigidly, but it should serve as a reminder that if you are going out on Thursday night for a basketball game, the study time has to be made up somewhere in the schedule.

Now that you have a schedule, where is an appropriate place to study? You should choose a place that is well lighted and reasonably quiet and where you will not be bothered by others. It is very difficult to study in your room if your roommate and others are having a discussion. You will be drawn into it and be distracted by it. This does not count as study time. Often the library will have small study areas that are quiet and where the distractions are minimal. Many students believe they must have a television, radio, or stereo playing while they study, but these usually create a diversion to their focus, especially television. Do not believe you can lie in bed and read something that requires concentration. Sit at a table or desk. Distracting sights and sounds take effort to ignore, and the energy and focus spend in ignoring the surroundings would be better expended in learning the material. This is why the section is labeled "minimizing your study time." Your goal is to learn material as quickly as possible, with as little repetition as possible, to free you for other interesting activities.

Part of what you study are notes from lectures. How do you record the lecture information? Your notes should not be a verbatim copying of what the instructor wrote on the board, but rather key words and phrases that when combined with the text material, clarify and expand upon the information in the text. This is one instance where being prepared for class will simplify the task of note taking, as you already know some of the important points and topics. The ability to take useful notes has to be learned. You must listen, comprehend, and take sufficient notes so that with later study you will remember what was discussed and why. It is important to go over your notes after class and amplify them while the material is still fresh in your mind; this is an excellent learning experience. Your notes are a valuable resource if they are legible and you can look at them five weeks later and know what was meant. If you cannot do this, then their purpose is not being met and their value is greatly diminished.

Your engineering course work consists of lectures; the responsibility for learning the subject matter is much more yours than it is in most high schools. (Courtesy of Hofstra University)

While the fundamental purpose of studying is to gain greater knowledge, a more immediate purpose is preparing for examinations on the material. This is your chance to demonstrate that you have assimilated the information of the course material and translated it into knowledge. The instructor's task is to assess whether and how well this translation has occurred. Cramming, staying up all night before the exam, is not the way to study for a test and is not at all the most efficient use of your time. There may be one situation when this type of study can be effective: when you must amass a large number of facts and be able by rote to repeat them on the examination; this does not occur in engineering or technical courses. Rather than deal with this narrow exception, let's analyze how you can best represent yourself to the instructor via an examination. Study the main topics of the course material, which you have done all along in small increments. Study the instructor, see how she presents the material, what aspects she considers important. Your preparation should certainly include these aspects. Pay attention in class; very often the instructor will indicate topics that may later appear on an examination. If you are too busy writing everything down in class, you will probably not be aware of these clues.

Ascertain what type of test it will be. If essay questions are asked, then you will direct your attention to summarizing important concepts, noting trends. You will want to prepare outlines of these concepts so you can form intelligent and thoughtful answers to questions about them. Should the examination be of the short-answer variety, the same information will be asked on the examination, but the relation between key words and specific information is stressed. Engineering courses often present problems to be solved on the

Your course work will require the use of computers. From word processing to equation solving, computers are integrated into the fabric of today's engineering curricula. (Courtesy of Hofstra University)

examination. Solving the homework problems and understanding their solution is the main way to prepare for this type of examination. Your homework must be sufficiently neat and detailed so that when you review it, the difficult steps or intuitive leaps stand out. You should examine these problems and understand the principles you are using. The principles are few, but the variables can be altered in many ways. For instance, in the case where there are three interrelated variables, knowing two allows for finding the third. Make sure you understand all the combinations possible; it is unlikely that the instructor will ask the same problem as in your homework, but with different numbers.

Remember that tests are also a learning experience, regardless of the grade. Usually we understand and learn from our mistakes, so the next time they will not occur, thus gaining a more fundamental understanding of the material.

REFERENCES

1. Brown, L., et al. *State of the World 1987.* Norton/Worldwatch Books, New York, 1987.
2. Brown, L., et al. *State of the World 1988.* Norton/Worldwatch Books, New York, 1988.
3. Calder, A. *Calder: An Autobiography with Pictures.* Pantheon Books, 1966.
4. Edwards, B. *Drawing on the Right Side of the Brain.* St. Martins, New York, 1979.
5. Florman, S. C. *The Existential Pleasures of Engineering.* St. Martins, New York, 1981.
6. Metz, L. D., and Klein, R. E. *Man and the Technological Society.* Prentice-Hall, Englewood Cliffs, NJ, 1973.

7. *The Technological Dimensions of International Competitiveness.* National Academy of Engineering, Washington, DC, 1988.
8. Waddington, C. H. *Behind Appearances: A Study of the Relations Between Painting and the Natural Sciences in This Century.* MIT Press, Cambridge, 1970.

PROBLEMS

1.1. Write a 200-to-300-word essay on why you are studying engineering.
1.2. Go to the library and, using an unabridged dictionary, copy the definition of the term *engineer*. Amplify this definition with your own ideas.
1.3. Prepare two different weekly schedules for yourself, examining different ways to use your time.
1.4. In your college catalog look at the course of study required for an engineering degree. Note the sequencing of courses in the natural sciences, mathematics, and engineering.
1.5. Make a list of the various student organizations on campus in which you might like to participate. You may need to contact people in student services to find out about these organizations.
1.6. In your hometown are there any technical/political issues such as waste disposal, nuclear power generation, or water and air pollution that are of concern? Do you see a role for a citizen-engineer relating to these issues? What is it?
1.7. Visit a local museum, or fine arts gallery, where sculpture is on display. Notice the structure from an engineering viewpoint as well as the creativity in the design. Most universities have many pieces of sculpture on display.
1.8. Investigate several engineering projects and determine if they can be completed without using science and mathematics.
1.9. Imagine you were Thomas Edison inventing the electric light bulb. What qualities would you try to create in your invention? Why?
1.10. Describe ten situations in your daily life that force you to be creative.
1.11. Imagine that in your hometown a decision has been made to install a monorail public transportation system between two major parts of town. Create the scenario of conflicts, technological as well as political, that might result.
1.12. Design a course of study for the kind of engineer and human being you would like to become.
1.13. Describe the ways in which being an engineer requires knowledge of political science, economics, and psychology.
1.14. Describe your method of studying in high school and consider how you might augment this to deal with the increased expectations of college.
1.15. Imagine the world without electricity. Name ten things that would be more difficult.

1.16. List five ways in which sports have been influenced by technology, for example, graphite-epoxy tennis rackets, metal baseball bats, and fiberglass boats.

1.17. Discuss the questions raised by the changing role of engineers who become business executives or entrepreneurs.

1.18. Discuss the many effects that engineers have had on water, such as through the creation of dams and water systems.

1.19. Describe the attributes of an ideal engineer.

1.20. Look at Alexander Calder's stabiles and mobiles and consider the engineering feat of his works.

1.21. Discuss the impact of electronics on rock music.

1.22. Describe a situation in which you developed a creative solution to a problem. Was your self-image enhanced by the experience?

CHAPTER 2

Historical Perspective

CHAPTER OBJECTIVES

To grasp the ramifications of technology on past lives and civilizations.

To learn the major transformations that have occurred in Western civilization.

To understand the coalescing of a variety of factors that created the industrial era.

To trace the formation of engineering professional societies and the development of the engineer's identity.

(W.C. White photo from The Library of Congress.)

Very often we tend to think of engineering in terms of the present day and fail to realize its importance throughout human existence. This chapter will help in changing this perception.

2.1 INTRODUCTION

Why should we examine engineering innovations from centuries ago when there is so much to learn about the present day? We can develop an understanding of the ramifications of technology on our lives today by seeing the effects of earlier technologies on the society of that time. Predicting effects is very difficult and questionable, whereas developing a sensitivity to potential effects is not as difficult and is something we hope engineers possess. There are several ways to study history, and one of the most interesting is to analyze the significant watershed events and determine what caused them. The underlying effects are usually fundamental, as when surplus food supply allows a population to increase, which in turn causes a restructuring of society. Engineering is always pivotally involved in the creation of new technologies, which in this case allows the food supply to increase and leads to the creation of additional technologies permitting societal reorganization.

Another aspect of historical change that we must remember, though it is difficult at times, is the time involved. We tend to look for change in times measured in days, perhaps years at the longest, whereas time in the historical sense is measured in decades and centuries. For this reason our deducing what effects current technological advances will produce is uncertain, but we can be certain that the advances of today will produce changes in how we live and how society is structured in the future.

2.2 PATTERNS OF CHANGE

There have been three major revolutions in the way society has been organized from the beginnings of humankind to the present day. For tens of thousands of years humans lived as hunter-gatherers; the first change occurred when they organized into agricultural communities, and agrarian societies formed the basis of civilization. The second change occurred with the evolution of these societies into industrially based ones. The third change is occurring now, as we move into a post-industrial based society, sometimes called a computer-based or information-based society. What has caused these reorganizations, and what important technologies were involved? We need to examine these questions to understand our roles as engineers, as good citizens, in shaping the society of our progeny.

Human beings tend not to change unless there is some force, external or internal, that directs them to do so. Societal organizations follow the same pattern. The hunter-gatherer groups did not want to change, but the force of hunger directed them to do so. In hunter-gatherer societies the people fed themselves by gathering wild fruits, nuts, and vegetables and hunting animals. Scarcity of food, caused by the end of the ice age ten thousand years ago, forced them to consider forming into agricultural communities, for better control of their food supply. This required an entire shift in the way society was organized; a nomadic tribe or group of people deals with problems differently than a village of people. A different governing structure is needed, different skills are required.

The agriculturally based societies flourished from about 3000 B.C. to A.D. 1300. By this time there were scarcities developing in food and energy supplies throughout Europe. The large feudal estates had produced a surplus of food, the population had expanded and started forming small villages and towns, but the population and the requisite food supply could not be sustained. Famines occurred, and starting in 1348 the Black Plague swept through Europe repeatedly for a hundred years. The effect of the plague was enormous, as millions of people died. In addition, wood was scarce, as more forests were cleared to create farmland. Wood provided the fundamental energy source for the population; it was burned for heating and used in every manner in forming utensils and implements found in daily life.

A new energy source, coal, had to be used, and a new societal organization, the industrial society, made use of the people who now lived in cities. Industries need concentrated population centers to draw upon. A new manner of understanding the world evolved as well, which set the basis for the scientific method of analysis. The Renaissance took place during this transition from the medieval era to the industrial age. It was a time of turmoil, change, and heightened creativity, not only in technology but in all areas of human activity.

The earth viewed from space. This gives us a sense of unity, globality, as well as the need to preserve the finite resources of our planet. (Courtesy of ASME)

What were the changes in thought, a fundamental shift in the philosophy of life, that occurred in this time of transition that allowed people to make the transition from a feudal, agrarian society to an industrial one? People credit Isaac Newton with discovering the laws of mechanics, which gave rise to our increased understanding of the physical world. He deserves credit for this, but the process started before him. The forces for the creation of the industrial age had been gathering strength for centuries. In 1620 Francis Bacon published *Novum Organum,* in which he espoused an objective way of thinking about the world. This is the way much of western society views the world now, but prior to Bacon scientists and philosophers were concerned with the why of nature. Why did this occur, what was the reason? This is often credited as being the Greek way of studying nature. Bacon on the other hand was concerned with the how of nature. Certain phenomena existed, why did not matter; how could they be explained, and with the explanation how could humans control them? This is a more directed approach and elevates our position as humans to controlling nature, rather than coexisting with it.

It is not sufficient to say this is the way we should think: there must be a structure that allows control to occur. René Descartes provided this structure. Following the publication of *Novum Organum,* though not necessarily because of it, Descartes concluded that it is possible to quantify all of nature as matter in motion and that mathematics holds the key to understanding. Mathematics is the structure needed for the "how" philosophy. Much of nature can be so categorized. You, for instance, have a certain size and exist at a certain point on the earth's surface. This can be developed into mathematical models; peoples' emotional sides cannot be so readily modeled, however. Cartesian coordinates, named after Descartes, are used extensively in defining where objects are at a given moment. At this point Isaac Newton entered and provided the means to use the mathematical models. With Newton's laws of motion describing movement in a gravitational field, it was possible to model mechanical systems and predict their behavior ahead of time—a tremendous advance. However, philosophy had shifted from coexistence to control, from seeing quality to seeing quantity.

We should not leave this discussion with the thought that only scientists and engineers were affected by a quantifiable world view. The change in attitude was reflected in economics and history as well. Adam Smith in his book *The Wealth of Nations* tried to develop a set of laws to parallel Newton's mechanical ones. His belief was that material self-interest was the driving force for all individual behavior as well as the behavior of the collective individual, the nation-state, and it is through Smith that the concept of laissez-faire was advanced. John Locke developed a similar view of the function of government. Thus, all facets of society undergo a shift in function and organization as a new structure emerges.

Part of the excitement of being alive now is that this too is a time of great change, a new renaissance with tremendous creative potentials at play and

2.3 BEGINNINGS OF ENGINEERING

with the possibility of a new societal structure forming that we can contribute to. Engineers have always been the creators of the tools that allow the new structures to form.

Let's examine a little more systematically and specifically the changes that occurred as we moved from a hunter-gatherer society, to an agrarian and finally an industrial society.

How can we say when engineering first began, as there were no records and engineering, as we conceive of it today, did not exist? We will assume that whenever there was an invention or innovation, engineering was required. Thus, engineering really begins with the first humans.

EGYPT AND MESOPOTAMIA

Humankind existed for tens of thousands of years, organized in small, primitive nomadic tribes and living by hunting animals and gathering fruits and vegetables. Figure 2.1 shows a hunter using a spear thrower, the result of early engineering. The spear thrower effectively lengthens the arm and gives a greater force to the thrown spear. About 10,000 B.C. the last ice age ended, causing the glaciers to retreat and the summer temperatures to rise. The increased temperature meant a decrease in rainfall and vegetation, and the

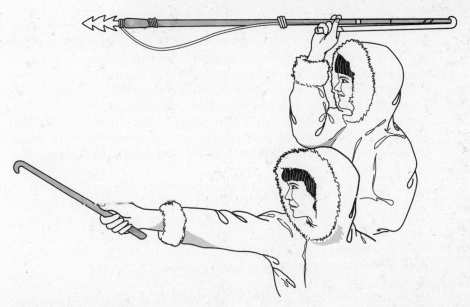

Figure 2.1 A spear thrower, used to increase the velocity of the spear by effectively extending the person's arm.

animal population in the high grasslands migrated to river valleys in search of water and food. Humans followed the animals into the valleys in northern India, Syria, Central America, and Egypt. Undoubtedly there are other places of early settlement as well, but the archaeological remains have not been discovered. In Egypt the nomads found the river valley of the Nile; in Syria the Tigris and Euphrates rivers were the lure. The geography of the rivers, their flooding times, and the fertility of the soil directed the type of society that would be formed.

Lest we jump too far ahead, the nomadic tribes had to develop rudimentary skills in farming to create an agricultural surplus to support the rest of society. The first farming was done by hand, using a pointed stick to till the soil. Eventually animals were domesticated to assist in farming, and a plow was invented. Figure 2.2 shows a scratch plow, which remained unchanged for thousands of years. A yoke had to be developed before the plow could effectively be pulled by oxen. Thus, the scratch plow, a seemingly ordinary technological achievement, allowed Egyptian and similar societies to feed themselves and increase in population.

The Nile is formed by two rivers, the White and Blue Niles, the former bringing decayed vegetation to the valley and the latter potash-rich soil. The annual flooding of the river valley occurred in the spring, so the rich soil could be planted and crops harvested in the summer. The population expanded because a surplus of food could be grown in the rich river valley.

There were not enough animal skins to clothe the increased population, so a new material was created: cloth, woven on a loom such as the one in Figure 2.3. The fiber was either animal hair such as wool, or a vegetable fiber such as flax, which is woven into linen.

The agricultural surplus was created because the scratch plow brought greater acreage into production than could have been farmed by humans alone. The surplus supported the craftspeople, the carpenters, potters, musicians, and bakers, as well as the administrators. The first administrators were probably those people whose knowledge of astronomy allowed them to predict the flooding of the valley, when to plant. Irrigation systems were developed to extend the arable land. Communities formed around certain land divisions, and in Egypt the communities were integrated into a kingdom, whereas in Mesopotamia the communities formed into independent city-

Figure 2.2 A scratch plow, fundamental to the expansion of an agrarian population. It allowed tillage of much larger acreage for crops than hoeing with a stick.

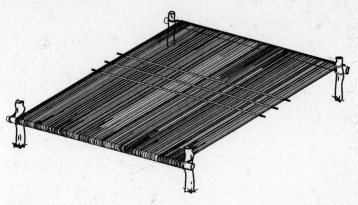

Figure 2.3 A loom used for weaving fabric, a replacement for animal skins, for articles of clothing.

states. In part because of the annual flooding of the communities, arithmetic and geometry were developed. Land had to be surveyed to locate boundary markers; areas were calculated, weights and measures standardized. Canal construction of the time likewise required development of the same analytic skills.

The tools of construction the Egyptians had were primitive by our standards; the lever, the inclined plane, and the wheel. With these tools, and, of course, with thousands of slaves the great pyramids that we associated with ancient Egypt were constructed. The Great Pyramid, with a square base of 756 feet per side, was built within one inch of square. The size is immense; the total number of blocks is about 2.5 million, each weighing two and one-half tons. The construction force has been estimated to have been a hundred thousand men during the flood season, and the construction project took 20 years to complete. The stone was transported from a quarry 500 miles away and probably required the year-round work of several thousand men. The organization, transportation, and feeding of this number of human beings for this long are staggering. It is only with the advent of the industrial age that the use of humans as the primary energy force in construction projects diminished.

This phenomenon was not restricted to the Middle East. The Great Wall of China, which is 3900 miles long including the various branches, though nominally covering a terrain of 1700 miles, was also constructed. The wall is 25 feet high and 20 feet wide, took 400 years to complete, and was started in 3 B.C., about the time of the Roman Empire. Untold thousands of slaves died and were buried in the wall during the construction process.

Thus far humans have had only wood, bone, or stone to work with; how did they make metal objects? Records from about 4000 B.C. indicate that rudimentary metallurgy was understood. Copper ore was reduced to copper metal by placing the ore in a charcoal pit fire. The copper metal melted, settled to the bottom of the pit, and was covered by a glassy slag. The slag was chipped away, and the copper was remelted in a crucible in another fire. The molten copper could be poured into molds. Figure 2.4 illustrates some molds from 3000 B.C. The molds were crude, and the metal had to be hammered to

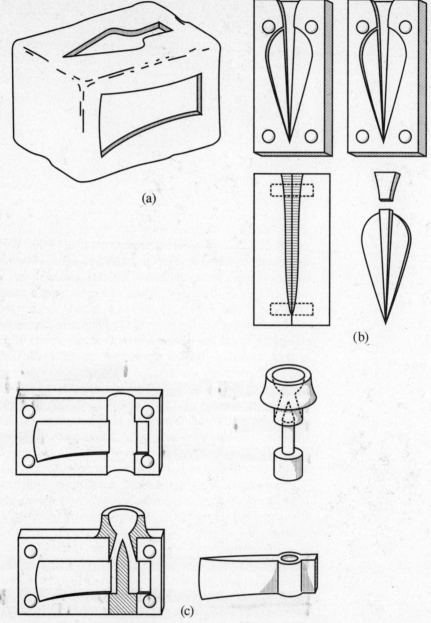

Figure 2.4 (a) A simple mold used to form an axehead in one indentation and a spear point in the other. (b) A two-piece mold used to form a spear point. (c) A false-core mold used to form an axehead.

its final shape. Note the progress from a simple mold in *a* to the more complex molds in *b* and *c*. By trial and error metallurgists of the era found that mixing some tin ore with the copper ore produces a more useful metal, bronze. Bronze has a lower melting point than copper and thus can be more easily worked, but is harder than copper. Bronze was first produced in northern

 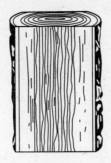 

Figure 2.5 How a three-piece wheel is made from a tree. Note that the inner wood, older and firmer than the outer wood, is used for the wheel.

Syria or Turkey, where the tin ore naturally occurred. It was about a thousand years before bronze appeared to any extent in Egypt.

We take the wheel for granted, but it does not occur naturally and had to be created. From about 3500 B.C. there is evidence of the two-wheeled cart, with wheels of wood. Figure 2.5 illustrates how a wheel was cut from a piece of timber. The outer new growth is cut away from the center hardwood. The remaining plank is divided as shown. Indications are that the cart was not used for long-distance transport, but for short hauls, as the axle was attached to the cart by leather straps, so the cart could be dismantled in a rough spot.

The primary scientific advances of the Egyptian and Mesopotamian era were in measurement, surveying, calculation of areas and volumes, and weights and measures. The Egyptians were able to devise a reasonably accurate calendar, probably through their knowledge of astronomy, and elementary use of the right triangle in land measurement. The Mesopotamians developed a method to solve simultaneous equations, though the purpose for learning it is not known. Advances in metallurgy included the development of the bronze alloy of copper and the use of molds in casting weapons, farm implements, and tools. In general the mathematical requirements for running the agrarian societies were not great, and the technologies that developed did not require mathematical skills as a basis for their implementation. Indeed, it was much later in human engineering history that such skills were needed.

GREECE AND ROME

The era from 1000 to 300 B.C. represents the development of the Greek culture as we are generally aware of it. Drawings from artifacts indicate that the knowledge of metallurgy was increasing to include ironwork; thus hand tools and weapons could be made stronger. The Greeks made advances in sailing and in ship construction and stability; however, the larger ships often were human powered by slave oarsmen. In pottery the Greeks made strides by raising the potter's wheel for greater control and developing an elaborate firing sequence for the kiln. They were able to fire to temperatures of 1000°C. The development of philosophy and of the aesthetic sense were of particular importance to the Greek culture. The Greeks reached conclusions regarding the form of the perfect rectangle; the long side is 1.6 times the short side (the

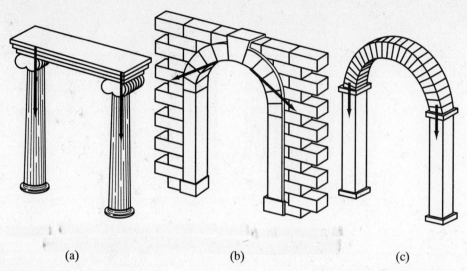

Figure 2.6 Diagrams of various arch types: (a) a pillar and lintel arch; (b) a Roman arch; (c) a true arch.

Golden Rectangle). They decided on the column shapes, lengths, and spacings to be used in building construction. This type of restrictive attitude did not lead towards technological innovation, nor was change viewed positively. The Greek view of the world was that it was created in its most perfect form by God. History then is the process of decay and increasing chaos, until finally God restores the world to its original condition and the process starts again. The ideal state was one that allowed few changes, slowing the process of disintegration.

Technological development was not great during the Roman era. Both the Greeks and Romans used the technologies of people they conquered and adapted them to their own use. The Romans were excellent civil engineers; some of their roads still exist. The roadways were constructed by laying a deep subbase of stone followed by a compact base. This method of construction allowed for slow wear, drainage of water, and no heaving of the road in cold weather. During the height of the Roman Empire 18,000 miles of roadway stretched from Turkey to Great Britain. These marvelous roadways were essential to the maintenance of communication between the colonies and the central government in Rome.

The Romans are also recognized for their use of aqueducts in providing fresh water to cities. The aqueducts ran for miles from the country lakes to the urban areas. This required great skill in measurement of distance and elevation to keep the water flowing. The Roman arch, or true arch, shown in Figure 2.6 supported the aqueducts and roadways. The figure shows the evolution from an arch requiring a buttress to hold the force exerted by the load on top of the arch. The true arch transmits this force vertically to the ground, eliminating the horizontal component and the need for the buttress, or side support.

A puzzling aspect of the Roman era is the unevenness of the technology. The Romans developed elaborate heating systems for buildings and developed

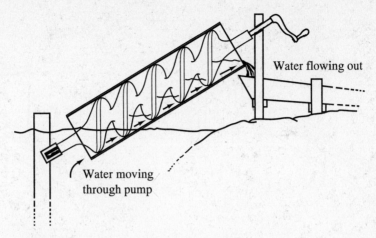

Figure 2.7 A screw pump used to raise water.

boilers so people could have hot water available on tap, but used a primitive bowl of oil with a wick for lighting. This bowl-lamp was used twelve thousand years before for lighting the caves of the nomads.

There were societal reasons for the constrained creativity of engineers of this period, concerning what an educated person should have as a career in Greek or Roman times. One could be a philosopher, politician and/or jurist, a general, or all of the above. An engineer could act to enhance the politician's or general's reputation. Thus, building roadways or water supplies and draining swamps increased the politician's standing, and designing better weapons increased the general's status. Mechanisms were developed to do just this. Significant creative ideas did occur, but the follow-through necessary to perfect them was missing. For instance, the Greek Archimedes devised the screw pump, shown in Figure 2.7, and the compound pulley used for lifting heavy objects. He also designed military engines for the defense of Syracuse. The screw pump raised water continuously and could be used in irrigation, in water supplies, and in draining water from mines. Of course humans had to supply the cranking power—the role of slaves dominates this time period. Releasing people, often slaves, from toil would create unrest, as society had no place for them, so making life easier was not a motivation. Most of the output of inventors was directed towards gimmicks, such as doors of temples opening mysteriously when a fire was lit. The fire would heat air, causing a volume change, and the expansion would, through gears and pulleys, cause the door to open.

The crossbow and catapult in Figure 2.8 were invented by the Greeks and improved upon by the Romans. The Romans also devised the waterwheel; Figure 2.9 illustrates a horizontal waterwheel and Figure 2.10 the more common overshot waterwheel. Despite its many inventions the Roman Empire did not last and eventually was overrun by what are called the barbarians from the north and east. The people were not necessarily barbarians: after all, which culture was addicted to slavery?

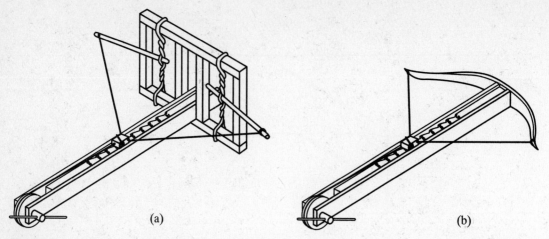

Figure 2.8 Sketches of (a) a catapult and (b) a crossbow.

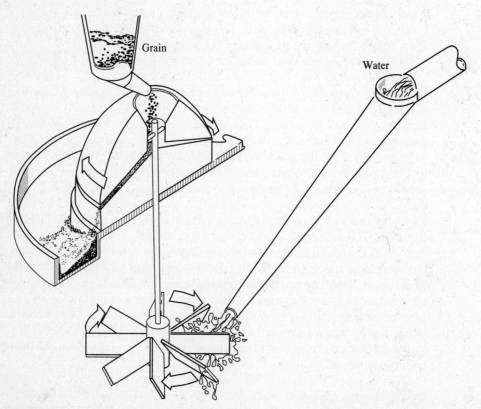

Figure 2.9 A waterwheel used for grinding grain.

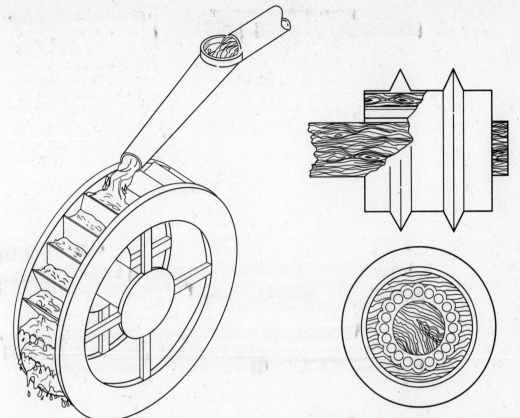

Figure 2.10 An overshot waterwheel.

Figure 2.11 A wood and bronze roller bearing used to support an axle.

These nomads from the north had independently developed technological advances of some sophistication. For example, Figure 2.11 illustrates a roller bearing that was used about 100 B.C. by people in Denmark. The technology did not last and may have died with the inventor, but it was a significant breakthrough in the support of a rotating shaft. Certainly the seamanship of the Gauls and their vessel design were superior to that of the Romans and was so noted by Caesar. All this is to show that technological inventiveness is not dependent on a civilized culture, but on a creative mind. An accepting culture is necessary for the implementation of new technologies and their improvement.

DARK AGES AND MIDDLE AGES

With the collapse of the Roman Empire in the fourth and fifth centuries A.D., what is known as the Dark Ages descended on Europe but not throughout the world. Perhaps the Dark Ages were not really so dark, as it was during this time that animals, and some waterwheels, began to replace humans as the power source. The Arabs were developing paper making, chemistry, and optics, and the Chinese were developing clocks, astronomical instruments, the loom and spinning wheel, and gunpowder. The importance of paper cannot be overestimated, as it allows for the written communication of ideas.

During this period the word *engineer* began to appear. Its root lies in the Latin word *ingeniare*, "to design or devise." About A.D. 200 *ingenium* was used to described a battering ram, and by A.D. 1200 *ingeniator* was used to describe a person who operated machines of war. Thus, there is a tie-in with inventiveness and warfare in the word and background of engineering. It remained for the industrial age to blossom with the use of mathematics and Newton's laws of motion, for the engineer to combine native inventiveness with scientific theory to create the power of design and analysis available today.

We notice that there has been a continuous association of engineering with domestic and military applications throughout human development. Sometimes we are troubled by the engineering profession's association with the military, and it is something we must each resolve. Seldom do people associate negative qualities with Leonardo da Vinci, the famous artist who painted *The Last Supper* and *Mona Lisa*, but who is perhaps equally well known as a military and civil engineer. He offered his services to princes of Italian city-states as an engineer, designing catapults as well as bridges and buildings. What remains most remarkable about him are his sketches of future engineering devices such as the machine gun, breech-loading cannon, tanks, helicopter, drawbridges, roller bearings, and the universal joint. The list extends to many more ingenious devices. (Notice *ingenious* has the same Latin root as *engineer*.) If you look at Leonardo's career, you can learn more about the engineering profession at the time.

Engineering was divided into two branches, military and civil. This separation persisted until the late 1880s. The funding of engineering works came principally from governments, including feudal kingdoms, which were interested in buildings and bridges and military defensive and offensive weaponry. This has not changed significantly in today's world; a significant portion of engineering design and effort is supported by government funding. The same national self-interest persists in governments today, hence the need for engineers to be employed by governments in much the same way Leonardo da Vinci was.

2.4 THE DEVELOPING INDUSTRIAL AGE

We learned of the forces in society that changed the agrarian structure that had existed since 3000 B.C. to an industrially based structure. This did not happen abruptly or with the contributions only of Bacon, Descartes, and Newton. Rather there were many contributors to new understanding of the physical world, and it is impossible to mention all of them. Assisting all of these contributors was the invention of the printing press and the dissemination of information it allowed. Johannes Gutenberg created the movable-type printing press in about 1454. The alphanumeric characters were cast in lead, and the printer selected the letters to compose words. This task was performed by hand until the early 1900s, when the Linotype machine was developed, a sort of typewriter that mechanically selects the characters and sets them in molds.

In the 1600s Galileo discovered that gravitational acceleration, hence the velocity a body achieves while falling, is independent of weight. In his study of motion, Galileo also concluded that the earth moves around the sun, in contradiction to the view of Aristotle, proposed centuries before. It was this assertion that caused the church to label him a heretic at the time.

The number of people contributing to the advance of science and technology increased each year, as the accomplishments of one fueled intellectual curiosity and competitiveness in others. A student of Galileo, Evangelista Torricelli, linked hydrostatics and dynamics and was responsible, with Blaise Pascal, for the development of the barometer. Robert Boyle discovered the expansion quality of air and the correlation between temperature, volume, and pressure. Robert Hooke discovered that a material lengthens in proportion to the force exerted on it, up to the elastic limit, and in compression it shortens in a similar fashion. Christiaan Huygens developed the spiral watch spring and the pendulum clock and, using his knowledge of the pendulum, was able to measure gravitational acceleration. Isaac Newton and Gottfried Wilhelm Leibniz independently developed differential calculus, essential to the mathematical analysis of most physical systems. In 1698 Thomas Savery patented a steam pump, used in the draining of mines, wherein the condensing steam on one side of a piston would create a vacuum and cause water to be pulled into the other side of the piston. On filling with steam again, the water would be forced out. The valves were opened and closed by hand, and the pump could complete five cycles a minute.

These forerunners and others set the stage for the next century and a half of development. Thomas Newcomen and then James Watt developed the steam engine. Watt's was a superior design, and it helped provide the power needed in textile mills, iron furnaces, rolling mills, and other industries. Other steam-powered machines and devices were created. The steam locomotive was developed once Watt's patent expired, as it required a horizontal steam engine that Watt viewed as impossible.

While on the topic of engines we must mention the forerunners of the automotive engine. The first was devised and built by Jean-Joseph Étienne Lenoir in 1860 using gas and air. In 1862 Beau de Rochas published an analysis of the processes required of a successful internal combustion engine, and in 1867 Nikolaus August Otto and Eugen Langen developed a gas engine that was more efficient and faster than Lenoir's. In 1876 they designed a four-stroke cycle engine, the precursor of today's automobile engine, using the cycle defined by Beau de Rochas.

The discovery and development of electrical engineering came later in history than its mechanical counterpart. Pieter van Musschenbroek of the city of Leyden in the Netherlands developed a device to hold a static electrical charge, now called the Leyden jar, in 1746. This is now called a capacitor, although it was given the name *condenser* by Alessandro Volta. The next step in development came in 1785 when Charles-Augustin de Coulomb developed what is known as Coulomb's law, that between two small electrically charged

This 1910 steam shovel appears much older to our eyes, though all the elements necessary for its operation remain the same today. (Courtesy of ASME)

spheres is a force of attraction or repulsion that is inversely proportional to their distance apart.

Two Italians discovered the principles of electrical conduction. Luigi Galvani in 1786 discovered the reaction a frog's muscle has to electrical stimulation, and in 1782 Alessandro Volta discovered the principals for creating an electric battery. The invention of the battery was a tremendous step, as it demonstrated conclusively that electrical conduction existed. Heretofore only static electricity was thought to exist.

By 1822 André-Marie Ampère had experimentally confirmed the flow of electrical current, leading to the science of electrodynamics. In 1831 Michael Faraday hit upon the means to generate electricity. He found that electrical current was induced in a wire by either increasing or decreasing the intensity of the surrounding magnetic field or by moving the conductor through the magnetic field. It is from these principals that an electric generator is constructed. Concurrently an American, Joseph Henry, discovered the same phenomena as Faraday, but he did not publish his results until 1832. In 1865 James Clerk Maxwell published a paper, "A Dynamic Theory of the Electromagnetic Field," in which he developed equations relating electrical conductivity, electric and magnetic fields, and mechanical force. He concluded, in part, that electromagnetic waves travel at the speed of light. This was confirmed 20 years later by Heinrich Rudolph Hertz. In 1897 Joseph John Thomson discovered the electron, supplying the first physical evidence of its existence in nature.

Before we tire of the list of inventors and discoverers let's mention just one more, Jagadis Chandra Bose, a Bengali engineer and scientist whose initial field of specialization was electrical engineering, but who is best known in the area of plant physiology. In 1895 in India he demonstrated the transmission of electric signals through space, independent of a wire. In 1896 Marconi was awarded a patent for the same achievement. Bose was invited to London in 1897 and lectured before the British Association for the Advancement of Science regarding his measurements of electromagnetic waves. Often the people we remember and who are credited with an invention, or a leap forward, had competitors who reached the goal first but were not recognized. Likewise certain societies have discovered a new technology, such as the manufacture of gunpowder by China or the design of the steam turbine in ancient Greece, but did not use it in the way we expect or it is used today.

Throughout the twentieth century the development of new technology has increased at an exponential rate. The creation of new materials has resulted in a variety of products that could not have been designed and manufactured without them. Nylon, the first synthetic fiber, was introduced by the du Pont Company in 1938 and revolutionized not only the apparel industry, but all areas of manufacturing from automotive tires to mechanical gears. In the early 1900s scientists first learned of nuclear energy; in 1942 engineers and scientists first controlled nuclear fission, which led to the development of nuclear power. Of course, today we are immersed in computer technology: microprocessors control many appliances and monitor the efficient operation and fuel adjustment of automobiles. Yet the first electronic

This realistic statue is not a fossilized professor, but rather a creation of J. Seward Johnson, Jr. This sculpture could not be completed without technological advances in thermoplastic resins, which impregnate the statue's clothing. (Courtesy of Hofstra University)

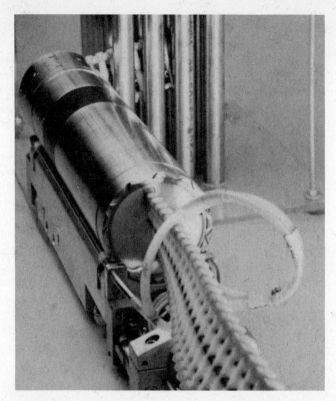

This strange-looking device is a robot, used to inspect and clean the inside surfaces of power plant components, such as steam generator tubes. This is particularly important in nuclear power plants where personnel are restricted from entering these areas unless there is a long shutdown period. A miniature camera attached at the end of inspection ribbon allows the operator to direct the robot's motion and observe the condition of the interior surfaces. (Courtesy of Public Service Electric and Gas Company)

computer was designed as recently as 1945. Also in the 1940s the transistor was invented and the development of solid-state electronics began. The list could go on and on, with jet engines, satellites, and radar, not to mention more mundane developments in building construction and automotive tires.

As engineers you will be leaders and creators in a new societal organization, dominated by computers and information flow. Computers will affect our lives as never before, especially when the energy crisis of the future arrives. Not only do computers assist us in many of life's daily tasks and work activities—they replace people. In manufacturing, machines can fabricate parts that used to require human assistance. In the area of information flow, computers permit a restructuring of industry; lower-level managers and workers can communicate with those at the highest levels, eliminating layers of middle management. This portends significant changes in career development patterns.

2.5 ENGINEERING SOCIETIES IN THE UNITED STATES

In Chapter 3 we will investigate engineering societies that exist today, but here we will look at the historical roots of these organizations. The development of these societies reflects the conflict or dual allegiance that engineers have towards their profession and their employer.

The roots go back to 1848 with the founding of the Boston Society of Civil Engineers and 1852 with the founding of the American Society of Civil Engineers, in reality the local New York section. Eventually, four other regional civil engineering societies were established, finally affiliating nationally under the auspices of the present American Society of Civil Engineers (ASCE). The purpose of these organizations was to advance the concerns of the engineering profession in all areas, technical and social.

Who were the engineers forming the societies? Many successful engineers were people who developed a business, perhaps a machine shop, into a sizable enterprise. We have noted the dilemma facing a person in such a situation: Are you an engineer or a businessperson? Where is your first identity? When the professional societies were formed, the same fundamental conflict existed concerning the functioning of the society: Should it be used to affirm the engineer's loyalty to an employer, that is, the business system, or should it direct the engineer to independent action safeguarding society? The result is that the societies pursue both aims, but this leads to conflict.

The high standards for membership, excluding many potential members, caused a more business-minded engineering society to form in 1871, the American Institute of Mining Engineers. As the engineering field became

The Robert Hoe Press and Saw Works Iron Foundry, from the 1850s, seems quite primitive compared to today's industry. (Courtesy of ASME)

more complex, it was no longer possible for one or two societies to address the needs of the diverse engineering community. Thus, the American Society of Mechanical Engineers was formed in 1880 and the American Institute of Electrical Engineers in 1884. Both of these organizations required meaningful standards to be met for membership and publication, but neither had standards as rigid as the ASCE's at the time, which created a hierarchical arrangement that precluded younger engineers from participation.

We can realize somewhat the changes the new engineers felt as they were the product formed by grafting the newly discovered sciences onto the millennia-old engineering craft tradition. Neither the organizations that were directed towards the craft nor the science societies were the right fit. The engineering professional societies formed to address this new constituency.

A small group of engineers in the early 1900s had a very positive feeling that the knowldege and abilities that were used in solving problems in the physical world—bridges, machines, and devices—could be adapted and used to solve social problems as well. Herbert Hoover was a leading advocate in this reform movement, and it caused some observers of the time, such as Thorstein Veblen, to portray engineers as the predestined leaders of social change. He wrote many books on economics and society, with his views on engineers reflected in his book *The Engineers and the Price System*. This reform movement united the fragmented engineering societies and perhaps can be credited with presaging scientific management, but it did not succeed in solving social problems.

Advances in technology, created by engineers, remove burden and danger in the workplace. The children pictured here were working in a textile mill in 1910. The affluence created by technology encouraged a more compassionate society that now prohibits child labor. (Courtesy of ASME)

Thus, engineering and engineers must be evaluated in light of the time they exist. We cannot hope to understand the engineers of the early 1900s if we do not concurrently understand the forces leading to World War I and the politics of the world following the war. Activism was part of the landscape in every nation: the Union of Soviet Socialist Republics experienced its birth, the class system in England was in disarray, the League of Nations was being created. This explains, in part, the need for as broad an education as possible in the social sciences and humanities, so that we may better understand the political, economic, and social environment around us.

Engineers are practical people and understand that businesses need to be financially viable to hire and employ engineers, accountants, machine operators, and office workers and that engineers play a unique and fundamental role in the organization. They create and design new products, which must operate correctly and safely, and at a cost that lets the company remain in business. All this is accomplished in the context of being a responsible professional, being responsible to the engineering profession and to society. It is in this area of societal responsibility, where the engineer has been active in the past, that renewed activity is needed in the future.

REFERENCES

1. Burke, J. *Connections.* Little, Brown, Boston, 1978.
2. Cardwell, D. S. L. *Turning Points in Western Technology.* Science History Publications, New York, 1972.
3. Hodges, H. *Technology in the Ancient World.* A. A. Knopf, New York, 1970.
4. Kirby, R. S.; Withington, S.; Darling, A. B.; and Kilgour, F. G. *Engineering in History.* McGraw-Hill, New York, 1956.
5. Kranzberg, M., and Pursell, C. W., Jr. *Technology in Western Civilization.* Oxford University Press, Oxford, 1967.
6. Layton, E. T., Jr. *The Revolt of the Engineers.* Case Western Reserve University Press, Cleveland, 1971.
7. Mumford, L. *Technics and Civilisation.* Routledge and Kegan Paul, London, 1962.
8. Rifkin, J. *Entropy.* Viking, New York, 1980.

PROBLEMS

2.1. List four examples of technological innovations that have affected the development of Western civilization; list four more affecting the United States.

2.2. Modern technology envelops today's world. Think about the modern supermarket: the door opens automatically, the food is checked out using a scanner. List ten more inventions that we consider everyday parts of life in the supermarket.

2.3 What were the principal reasons that Greek science did not lead to technological advancements?

2.4. Write a brief report on the construction of the Great Wall of China, giving an estimate of material usage and people required.

2.5. Paper is fundamental to our communication of knowledge. Prepare a report on paper manufactured from earliest times.

2.6. Describe Leonardo da Vinci's version of the helicopter and two other inventions of his that you believe are most significant in terms of today's society.

2.7. The first transatlantic cable was laid in 1858. Investigate the technological problems that first had to be solved.

2.8. Write a report on the development of the ceramic automotive engine.

2.9. Investigate and write a paper on the technological innovations that may be possible in light of superconductivity.

2.10. Some of the iron railroad bridges that were built in the mid-1800s failed, resulting in loss of life and property. Report on the causes of these early failures.

2.11. The Brooklyn Bridge was designed and constructed under the supervision of John Roebling and his son, Washington. Write a report on the design innovations and the problems they addressed.

2.12. Write a brief report on the discovery of semiconductors and breakthrough investigations of their properties.

2.13. Write a short biography of one of the following, pointing out technical and/or scientific achievements: Archimedes, William Gilbert, Otto von Guericke, Robert Watson, James Watt, Thomas Edison, George Westinghouse, John Gorrie, Robert Goddard, Alessandro Volta.

2.14. Describe how engineering has improved our lives in the areas of transportation, food processing, and medicine.

CHAPTER 3

Fields of Engineering

CHAPTER OBJECTIVES

To learn about the technological team.

To investigate the structure and requirements of engineering professional societies.

To find out what accreditation of your curriculum means.

To distinguish between major fields of engineering study.

To become aware of various technical career possibilities within any given field.

(Photo courtesy of NASA.)

In this chapter we will look at some of the more prominent engineering societies, at the types of careers members may pursue, and at the curricula of some typical engineering programs. In addition we will see what it means to be a professional engineer.

3.1 THE TECHNOLOGICAL TEAM

As a professional in the field of engineering you will work with other technically educated people to form what is known as a technological team. The team will bring complementary skills together to solve problems. In the broadest sense it consists of the following groups:

- scientists
- engineers
- technologists
- technicians
- craftspeople

The engineering graduate will come to rely on the knowledge and skills of others in the technological team, just as they will come to rely on his or hers. The education and training for each group of people is different but often overlaps in adjacent areas. Whereas the engineer and scientist both must be well educated in mathematics and science, the scientist's role is to seek out new fundamental understandings of the world, expanding our existing knowledge. A scientist seeks to understand the why of nature. However, very often scientists employed by companies are applied scientists, using the knowledge gained by advances in theoretical understanding to build new products or devices. This is what an engineer does as well.

The technological team usually is directed by an engineer, as the engineering function primarily involves the design of a new product, project, or system. The purpose of an engineer's education is to equip creative minds with the mathematical and analytic skills necessary to conceive new designs, to intelligently question present ways of accomplishing tasks, and to find better alternative methods, in light of evolving technology.

Whereas the scientist and engineer overlap functions on the abstract end of the spectrum, the engineer and the technologist may overlap at the implementation level. A technologist has graduated from a four-year engineering technology program. These programs typically have fewer mathematical requirements than an engineering program and courses more related to current hardware and processes. The presentation of material is less theoretical than

Field engineers are taking core samples from a road to determine the subsurface condition. (Courtesy of Sidney B. Bowne and Son)

in engineering programs, though the types of courses offered are quite similar. The technologist works with the engineer in all areas, often in the interface between the engineer and the technician, overseeing the technicians' work and further developing the design changes.

Once you are hired by a company, your initiative and abilities determine what assignments will be given you and what your function will be in actuality. The functions of the engineer and the technologist overlap in the workplace, blurring the lines of distinction education delineates.

The technicians are usually graduates of a two-year program in engineering technology. They are "hands-on" people and could be draftsmen, field inspectors on construction projects, or electronic technicians. Craftspeople are skilled workers, historically the first members of the engineering technological team. They have the skills to make devices and systems. Modelmakers are used in the aerospace industry, machinists in a variety of industries, welders and carpenters in the construction industry. Usually they have training in their skill, followed by years of practice.

Being an effective member of the technological team is important to your company and your career. College is the place to develop skills in team building. In laboratory experiments you may be one of a team of students performing the experiment. Be certain to assume your responsibility and become an active participant, not merely a passive observer. Extracurricular activities provide ways to engage in team building, through intramural or varsity sports, engineering student clubs or involvement in student governance.

3.2 ENGINEERING SOCIETIES

In Chapter 2 we saw that by the end of the 1800s the fields of engineering had expanded greatly, resulting in the formation of several engineering societies. This rapid growth in engineering fields has continued to the present day, resulting in over 400 engineering societies and related groups in the United States and 21 engineering societies in Canada. This should indicate to you, at least preliminarily, the tremendous diversity in career opportunities that exists.

Several organizations are concerned with the engineering profession as a whole, including the National Society of Professional Engineers (NSPE), the National Academy of Engineering (NAE), and the American Association of Engineering Societies (AAES). The AAES comprises 28 engineering societies and acts to represent the profession, for example in the area of governmental policies and through one of its subgroups, the Engineering Manpower Commission, which undertakes many studies regarding the engineering profession. These studies include research on industrial demand for engineers in the future and the current placement of engineering graduates, and a comprehensive salary survey. The AAES uses this information to support public policies necessary for the continual improvement of the engineering profession.

The beginnings of the unity of purpose hark back to the founding of the United Engineering Trustees, Inc., in 1904 by five engineering societies, called the Founder Societies. They are listed here with their dates of founding.

These research engineers are developing a sustained fusion reaction at the Tokamak Fusion Test Reactor at the Princeton Plasma Physics Laboratory. (Courtesy of Public Service Electric and Gas)

American Society of Civil Engineers (ASCE 1852)
American Institute of Mining, Metallurgical, and Petroleum Engineers (AIME 1871)
American Society of Mechanical Engineers (ASME 1880)
Institute of Electrical and Electronic Engineers (IEEE 1884)
American Institute of Chemical Engineers (AIChE) 1908)

The United Engineering Trustees is the titleholder to the 20-story United Engineering Center which provides office space in New York City for 23 engineering societies and related organizations. In addition, the Trustees provide to the profession the Engineering Societies Library and related information services as well as the Engineering Foundation, an endowed foundation dedicated to research in science and engineering directed towards the good of the engineering profession and humanity.

These 23 societies establish goals and disseminate information about themselves. Their objectives are similar to those established by ASCE, listed here.

1. To encourage and publicize discoveries and new techniques throughout the profession.
2. To afford professional associations and develop professional consciousness among civil engineering students.
3. To further research, design, and construction procedures in specialized fields of civil engineering.
4. To give special attention to the professional and economic aspects of the practice of engineering.
5. To enhance the standing of engineers.
6. To maintain and improve standards of engineering education.
7. To bring engineers together for the exchange of information and ideas.

The societies implement these objectives in a variety of ways. At national and regional meetings each society addresses these issues. Often at these meetings professional articles are presented by investigators in the field. A variety of technical journals sponsored by the societies disseminate technical information, and there are student chapters of these societies which you should join now.

There are several reasons for becoming a member while being a student. Contacts with engineers working in industry may be possible. You can participate in running the student chapter, creating an opportunity to speak in public and work as a technical team member. In addition you will receive a society periodical relating what is currently happening in the profession, intriguing you with future opportunities for involvement. An important benefit is that the societies all have loan assistance and scholarship programs, which you become eligible for. Following graduation your continued participation is necessary, so that the engineering profession can continue to advance. One of the ways that professional societies assist their membership is through continuing education courses. Although engineers are not required to pursue continued study, it is very beneficial to their careers and society if they do. Professional societies frequently offer short courses on new technologies so engineers can remain current in their fields and/or broaden their knowledge. Colleges and universities also offer specialized short courses and in some companies, consultants teach in-house workshops. We have seen that advances in engineering parallel advances in society; without your professional involvement advances occur more slowly. One method of being involved with technology utilization in society is through your professional society.

Another advantage of being a student member is that it's fun! Very often there are challenging competitions, such as the Rube Goldberg contest, against teams from other colleges. In addition, student members have many opportunities to meet practicing members which is interesting, educational and provides the potential for networking.

In addition to having worthy objectives such as those mentioned, the societies have membership requirements to assure that competent engineers direct the organization. The following is a list of the various grades of membership that ASME has and that are similar to those in the other societies.

STUDENT MEMBER. Belongs to a student chapter at a school with a curriculum approved by the Accreditation Board for Engineering and Technology (ABET).

ASSOCIATE MEMBER. Must have graduated from an engineering school of recognized standing, or have eight years of acceptable experience.

MEMBER. Requires six years of practice if one graduated from an engineering school of recognized standing; otherwise, requires twelve years. Must have spent five years in responsible charge of work.

FELLOW. Requires nomination by fellow members for an engineer or engineering teacher of acknowledged attainments. Also requires 25 years of practice, and 13 years in member grade. This is an honorary grade.

HONORARY MEMBER. Requires nomination by membership and election by the board of directors. This grade is given to people of distinctive engineering accomplishment.

EXECUTIVE AFFILIATE. Not necessarily an engineer, but in a position of policy-making authority relating to engineering; must be one who cooperates closely with engineers.

AFFILIATE. Capable and interested in rendering service to the field of engineering.

The position you can consider at this point is student member; all societies make the cost of membership minimal, and we have already examined the benefits. One additional benefit to being a student member is that you can automatically become an associate member following graduation. There is no membership fee involved, which there is if you do not choose this path, and you do not have to obtain membership forms and be nominated.

In addition to societies affiliated with virtually every field of engineering there are professional societies organized to support women and minorities. These include the Society of Women Engineers (SWE), the National Society of Black Engineers (NSBE), and the National Society of Hispanic Engineers (NSHE). All these organizations provide special scholarships and loans, provide role models, networking opportunities, and aid in specific problems that these groups may encounter in school and industry.

3.3 ACCREDITATION BOARD FOR ENGINEERING AND TECHNOLOGY (ABET)

One of the requirements for ASME student membership and for most others is that the university engineering program be ABET accredited. What does this mean? ABET's function is the examination and accreditation of engineering and technology curricula. ABET comprises 19 participating organizations and 4 affiliate organizations.

There are 31 program areas that ABET accredits for periods of three or six years. At the end of these periods a fact-finding team from ABET (Engineering Accreditation Commission [EAC] for engineering programs and Technology Accreditation Commission [TAC] for technology programs), comprised of members of the participating and affiliate professional societies, visits the university and analyzes the course of study for the individual programs. The team examines the quality of course work, the competence of the faculty, and the administrative support and, importantly, interviews students. The team writes a report of its observations and submits it to ABET. Based on this and other factors a decision to accredit and for what time period, or not to accredit at all, is made by ABET, and the school is informed of the accreditation action. This assures you, as a student, that the course of study you are pursuing is relevant, that it meets minimum professional standards, and that when you graduate you will be an asset to the engineering profession and to society. As of 1988 there were 1401 accredited bachelor's degree engineering programs, 292 accredited four-year technology programs, and 468 accredited two-year technology programs. Undoubtedly you are enrolled as a major in one of these programs presently. What follows is an overview of some of the fields of engineering you may pursue upon graduation and the types of projects you can be involved with. In Chapter 4, professional registration, licensure as a Professional Engineer, is examined in some detail. Graduating from an ABET accredited program makes the path to engineering registration significantly easier than graduating from a non-ABET accredited program.

3.4 FIELDS OF ENGINEERING

Often freshmen engineering students do not know what field of engineering they wish to major in. Others think they know and then, with more information, change majors. This is perfectly normal and reasonable. The following descriptions provide a brief overview of engineering fields of study. Talk to faculty members in various fields of engineering, read engineering society magazines (found in your college library), talk to practicing engineers if possible. In selecting a major, pay attention to your interests and abilities, do not focus on curent salaries and market demands which fluctuate over time.

AERONAUTICAL AND AEROSPACE ENGINEERING

The term *aerospace engineering* is often used to denote the entire field and the aerospace industry, but we should make a distinction between aeronautical and aerospace engineering. Aeronautical engineering deals with flight and the movement of fluids in the earth's atmosphere. Thus, the research, design, and development of all types of airplanes are included in this field. You will specialize in work areas centered on aerodynamics, propulsion, controls, or structures. One of the interesting aspects of aircraft design is the interaction between the air around the plane and the plane's structure: one influences the behavior of the other. If you have been in an airplane, you must have noticed that the wings flex as the airplane flies. They are designed to flex, and their movement affects the air flow around the wing. Furthermore, the development of new materials is important, as materials fatigue after too much flexing and the wings must be replaced. The more durable the material, the greater the wing's life expectancy. Notice the word *material*, not *metal;* much of the materials used in new aircraft design are not metals, but composites of graphite and epoxy.

As we still imagine ourselves aboard the aircraft, someone had to design our aircraft's engines, and has to redesign them as new materials allow different and better operating conditions. For instance, the inlet temperature to the jet's turbine can increase because new turbine blades are able to withstand continuous operation at higher temperatures. Not only is the turbine redesigned for the new conditions, but related equipment is redesigned to operate with a hotter fluid.

While we look to the stars and imagine people rocketing through space, aerospace engineers are required to turn this image into a reality. Quite similar to aeronautial engineering in the job aspects involved, aerospace engineering deals with environments not found on earth—space's vacuum or the atmospheres surrounding other planets. Aerospace engineers must design rockets to leave and reenter earth's atmosphere, structures to exist in space, life-support systems enabling humans to live in space, robotic devices, and control instrumentation. Let's consider just one application for an aerospace engineer, the problem of vibration in an orbiting space station. We seldom think about this on earth, even if the structure is ultralight, as the space station must be, because air assists in the damping of the vibrations. Thus, structures are passively damped, and no vibrations continue for long. (There are exceptions, such as the huge World Trade Center, in New York City, which has a mechanical damping system.) The structures in space must be actively damped; a control system must sense when a vibration is occurring and set up a countermovement in the structure to cancel the vibration, much as a reflected wave will cancel an incident wave.

Perhaps nothing has illustrated a high point in U.S. technological accomplishment as well as the landing of an astronaut on the moon, July 20, 1969. (Courtesy of NASA)

AGRICULTURAL ENGINEERING

For many of us located in urban and suburban communities, the concept of how food is produced is not one we dwell upon, but fortunately for us the agricultural engineer does. A problem since the beginning of humanity is the production of enough food to support the population. Being from an industrialized nation, we tend not to think of this problem, but on a global scale starvation is a reality millions must confront daily. The agricultural engineer blends engineering knowledge with understanding of soil systems, land management, waste management, and environment control to create methods and technologies that will allow the continuation of high crop yields to feed us. There are five general areas that agricultural engineers specialize in. Soils and water is the first and concerns itself with water drainage, erosion control, irrigation systems, and land use. The second area is food engineering, where engineers analyze food-processing procedures, finding ways to minimize waste, energy, and damage. Drying of foods would fall under this category; perhaps if you are a backpacker you have enjoyed the benefits of this technology. In this process the food is placed in a vacuum chamber where the water evaporates; then the food is sealed in an airtight package to prevent spoilage. You can easily imagine how many steps the process takes, all of which must be considered in the engineering design. Agricultural engineers are also involved with irradiation of food as a means of long-term safe storage. A third area is power machinery, the development of new agricultural feed systems and handling and processing machinery. The fourth area deals with structures, particularly those necessary for the housing of livestock and the food and waste-handling aspects necessary for this type of building. Last is the area of electric power utilization. In remote areas farmers and ranchers must generate their own electricity. The wise and efficient use of electricity is always important, particularly so in self-reliant situations.

ARCHITECTURAL ENGINEERING

The architectural engineer works with architects in the design of buildings, focusing on the analysis and design of materials used in construction. The architect in this situation is concerned with aesthetic requirements of the space and the environment it surrounds and surrounding it, while the architectural engineer's concerns focus on the structural integrity and safety of the design. There is a great deal of similarity between this and structural engineering, the main difference lying in the architectural engineer's concern for aesthetic considerations.

AUTOMOTIVE ENGINEERING

One of the major industrial forces in the economy of the United States is the automobile industry, and engineers are a vital component of its growth and strength. Automotive engineering transcends automobiles and includes all types of vehicles from trucks, to bulldozers, to motocycles. The field is an interdisciplinary one, as you can readily imagine by envisioning all the systems in today's automobile. There is, of course, the design of the engine itself, with its thermal and mechanical aspects, as well as the fuel and lubricant considerations. The structural design must be able to withstand impact and protect the driver and passengers. Here too the structural design depends on the materials used, with thermoplastics replacing metal in many instances, and on the aesthetic requirements so fundamental to marketing the car.

Testing an automotive engine system may require the use of a spectrum and network analyzer. (Courtesy of Hewlett-Packard)

BIOMEDICAL ENGINEERING

Biomedical engineering is a comparatively new field of engineering that began in the 1940s but, in a sense, has been practiced for thousands of years since the creation of the first artificial limb. The field is interdisciplinary, and engineers must be able to work with biologists and physicians in developing new equipment and materials. This means being able to communicate in their language as well as in the language of the engineering community. The three general divisions of biomedical engineering have names that are frequently used interchangeably. Bioengineering, a research activity, applies engineering techiques to biological systems. Medical engineering develops medical instrumentation, artificial organs, prosthetic devices, and materials. Clinical engineering concerns itself with the hospital system problems, such as decontaminating air lines, removing anesthetic gases from operating rooms, and the correct operation of instrumentation in health facilities. You would certainly want the monitor that a physician was using on a patient to accurately reflect the condition of the person. Very often this field requires a graduate degree because of the diverse nature of the material that must be comprehended.

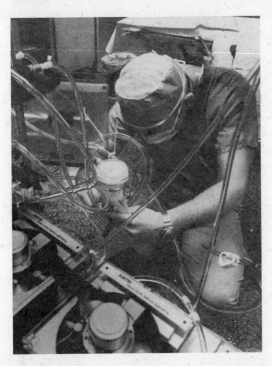

A technologist is installing a filter in a cardiopulmonary bypass circuit—a system design by bioengineers. (Courtesy of Pall Corporations)

CERAMIC ENGINEERING

When we think of ceramics, the image conjured up may be a green frog figurine, a bowl, or a pitcher. Whereas these objects are called ceramics, the field of ceramic engineering is indeed something far different and very interesting. Ceramic materials are nonmetallic, inorganic materials, such as silicon dioxide (a form of sand), that fuse at high temperatures to form a variety

A PC-based laser-interferometer is used in the calibration of machine tools, detecting small errors in machine motion. (Courtesy of Hewlett-Packard)

of materials. Space shuttles have much of their surfaces covered with ceramic tiles to withstand the high temperatures of reentry into the earth's atmosphere. Ceramic materials can withstand high temperatures and not lose their strength and, because they are chemically stable, can be used in corrosive and chemically active situations without deterioration. Rocket nozzles are made of ceramic materials, as are spark plugs. Ceramic engineers are needed to determine the material requirements for the particular application. For instance, diesel and gasoline engines have some of their metal surfaces coated with a thin layer of ceramic material to improve efficiency. The piston crown and exhaust valve seats are also coated and therefore can withstand a greater temperature. Also, the engine loses less heat to the cooling water, due to the insulating aspects of the ceramic coating; this allows the engine to convert more of the fuel's chemical energy into work. A ceramic engineer had to design the ceramic material to withstand the vibration of the engine and the thermal expansion of the metal in this situation. So ceramic materials do have some flexibility, even though we think of them as brittle.

CHEMICAL ENGINEERING

Chemical engineers translate the laboratory developments of the chemist and physicist into commercial realities. The chemical engineer works in the pharmaceutical, chemical, pollution control, nuclear; and electronics industries. In the electronics industry, for instance, chemical engineers must confront and solve the production problems encountered in the manufacturing process. The most commonly used method to manufacture electronic components is evaporating and redepositing a material on a surface a distance from it in a vacuum chamber. The substance is housed in a vacuum chamber and has a large amount of electrical current passed through it. The substance evaporates, with its atoms traveling to a surface about one foot away and deposited there to a thickness of less than one micron. Remember that a micron is one-millionth of a meter or one thousandth of a millimeter. This process must be accurately controlled and measured—a job for a chemical engineer.

Chemical engineers are now involved in the pharmaceutical industry where a new antibiotic, Xmycin, has been developed by the biochemists in the research laboratories. The next stage is to design a pilot plant to manufacture Xmycin. Problems of scaling up the processes must be analyzed and overcome, and the equipment must be designed and controlled. Once the manufacture of the antibiotic is successfully accomplished in the pilot plant setting, the company can decide whether or not to proceed to a full-scale manufacturing facility. Chemical engineers must redesign the equipment for this situation.

Chemical engineers are involved with all types of processing equipment and manufacturing, from petroleum products, to paints, to nuclear fuels, to vitamins. All of these plants have chemical processes to control and equipment that must be designed for the characteristics of the substances under consideration.

These electronic circuit boards are cleaned with extremely pure water. This technology, developed by chemical engineers, is environmentally preferable to one using fluorinated hydrocarbons. (Courtesy of Pall Corporation)

CIVIL ENGINEERING

The field of civil engineering is one of the most apparent to us. The highways and bridges we drive on, the tunnels we drive through, the buildings we live and work in, and the water supply and sewage treatment systems we depend on are the handiwork of civil engineers. As civil engineers you need to understand a variety of situations that go beyond engineering mechanics. To be sure, part of your course of study and career opportunities lies in the construction industry, structural design, and overseeing the construction process. Buildings rest on the ground, and you must understand and be able to determine the load-bearing properties of soil and, for instance, how to increase the load properties with the use of pilings.

The construction of highways requires accurate surveying of the terrain, showing exactly where the roadway is going and defining aspects of the terrain that need to be considered in the highway construction. A highway is more than the construction of the roadbed: the engineer must consider the effect of drainage, land use and environmental impact, and the interconnectedness with other transportation systems. Bridges are one of the engineering achievements that most strongly capture our imagination, such as the Golden Gate Bridge in San Francisco or the Verrazano Bridge in New York City.

Another important area of civil engineering is waste treatment, including garbage, sewage, heavy metals, and toxic materials. We personally create a tremendous amount of waste in the course of our daily lives. In 1985 the household waste generated in the United States was 136 million metric tons.

Fundamental to the design of structures and roadways by civil engineers is a land survey. (Courtesy of Sidney B. Bowne and Son)

Most surveys require line-of-sight measurements. However, a global positioning system allows engineers to determine a three-dimensional position on the earth's surface using satellite communication. This highly accurate method can be used effectively in urban areas, even where buildings block the line of sight, and in geographically difficult regions, such as hilly terrain and canyons. (Courtesy of Sidney B. Bowne and Son)

(A metric ton is 2200 pounds.) This does not include waste from the industrial sector. How to dispose of this is one of the major problems facing municipalities today, and civil engineers can help design the necessary systems.

Because what civil engineers construct is so vital to society, most civil engineers are licensed professional engineers, more so than engineers in any other engineering field. Also, because of their involvement with large-scale municipal systems, many of the engineers employed by local and state governments are civil engineers. Over 50 percent of the city planners in the United States have a background in civil engineering.

COMPUTER ENGINEERING AND COMPUTER SCIENCE

Computer engineering is closely connected to computer science, and while most of computer science is not an engineering field, students often ask whether they should study computer engineering or computer science. Both are involved with the design and organization of computers: its software and hardware. Computer science approaches this from the software viewpoint with less emphasis on the hardware, and computer engineering from the hardware viewpoint with less emphasis on the software. Computer science in the most general sense is the study of problem-solving procedures, computability, and computation systems. Calculus, which is based on the mathematical differential, is not as relevant to computer science as finite mathematics, because digital computers deal with discrete pieces of information. The field of numerical analysis concerns itself with approximate solutions of sets of equations on digital computers. Software engineering deals with the creation of software to solve problems.

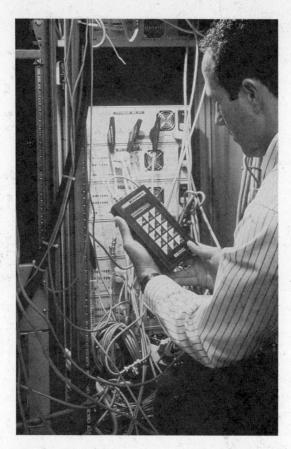

A hand-held scanner can instantly locate local area network (LAN) cabling problems. (Courtesy of Hewlett-Packard)

ELECTRICAL AND ELECTRONICS ENGINEERING

Electronics engineering is often called electrical engineering: however, electronics engineering excludes large electrical systems, such as found in motors, generators, electrical circuits of buildings, and power transmission systems. Electronics engineering deals with the passage of charged particles a gas, vacuum, or semiconductor. Electrical engineering, exclusive of electronics, deals with the motion of electrons in metals, such as through wires and filaments.

This description does little to show the tremendous variety of ways in which electrical engineers are involved in today's society. A blackout dramatizes how dependent we are on systems with electrical components. Electric power eases our daily life tremendously. The typical U.S. household has at least ten electric motors, and this number can easily be multiplied several times for some homes. Every fan, dishwasher, washing machine, and refrigerator relies on motors for its operation. Automobiles are also equipped with a variety of electrical and electronic devices, from the fundamental battery to sophisticated electronic fuel injection systems. All are designed by teams that include electrical engineers.

3.4 FIELDS OF ENGINEERING **61**

Electrical Engineers designed a rugged and portable transceiver test system that can be used in remote locations. (Courtesy of Hewlett-Packard)

The growth of electrical engineering in the last 40 years has been exponential. Microprocessor-based control systems are used in most major home appliances, such as microwave ovens and dishwashers. Computer-based control systems are standard equipment in aerospace systems. Control systems provide rudimentary intelligence to devices, so they can sequence operations. Some modern aircraft are unstable unless computer-controlled; pilots cannot

Electrical engineers created this digitizing oscilloscope. In this situation the oscilloscope is being used to measure hard-to-find circuitry turn-on problems. (Couryesy of Hewlett-Packard)

respond quickly enough to fly forward swept wing planes without computer assistance.

In experimental activities instrumentation frequently relies on an electrical (electronic) signal to show that a certain event has occurred. For instance, thermocouples measure in millivolts a voltage potential that correlates directly with temperature: knowing the voltage allows determination of the temperature.

It is difficult to imagine an engineering discipline that does not include or require the use of electrical engineering at some point in the system design. Let's consider a lesser known area of electrical engineering, oceanic engineering. Electrical engineers are developing technology and analysis methods to determine the sea state of the ocean through satellite observations. The satellite emits and receives an electrical signal that is distorted by the ocean's sea state, the wave height and velocity, and wind-generated surface conditions. By knowing what the weather conditions are in remote ocean areas, ships can avoid turbulent weather, and weather forecasts can be more accurate.

We are in what some people call the communication age, the age of information overflow. How is this information created and transmitted? By electrical and electronic devices. Computers generate information from programs they run; the information may be transmitted across country by satellite and through optical fibers, or stored on diskettes.

Every piece of electronic or electrical equipment requires the knowledge that an electrical engineer possesses to be designed, developed, and manufactured.

ENVIRONMENTAL ENGINEERING

This specialization in engineering has traditionally been affiliated with the civil engineering programs at most universities. It is, however, developing a separate identity and crosses the boundaries of many disciplines. All engineers are concerned about the environment and in creating processes and products that minimally disrupt the natural environment. Thus, environmental engineers may be chemical engineers focusing on the containment of environmentally hazardous materials within a plant, preventing their release. They may be mechanical engineers concerned about air pollution caused by combustion processes, or civil engineers looking at waste disposal problems and water quality issues. Environmental assessment of manufacturing plants, proposed and existing, is very important in the public eye, and engineers are creating plants, facilities, and systems that minimize adverse environmental effects during normal operation and in cases of natural disasters.

INDUSTRIAL ENGINEERING

Industrial engineers are concerned with the design, improvement, and installation of integrated systems of people, materials, and energy. What are examples of these systems? Not all systems have to have all components. Consider

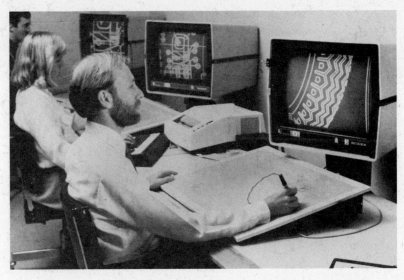
Mechanical engineers designing a tire mold. (Courtesy of Michelin Corporation)

the manufacture of tires: a machine is operated by a person and needs certain raw materials to produce the tire. The problem for the industrial engineer is to bring the material to the machine as it is needed, remove the finished product, and transport it to the next workstation. When establishing a manufacturing procedure, the industrial engineer considers the machine's tire capacity, when the machine needs routine maintenance, and what the machine operator should do when the machine is not functioning. Industrial engineers are always analyzing manufacturing processes to determine if the sequence of operations can be made more efficient so the manufacturing process is less costly per tire, making the company more competitive, while maintaining or improving the tire quality. The role of the industrial engineer has increased and is critical to a company's financial success, as the technologies that must be integrated are increasingly complicated and costly, and the competition in the world marketplace is great.

Your background as an industrial engineer will include areas in management, manufacturing, plant design, quality control, data processing, and systems analysis. Industrial engineers must be people-oriented and will find themselves involved with incentive programs, employee benefits, subcontracting and contracting, and areas of financial management.

MANUFACTURING ENGINEERING

Manufacturing engineers are concerned with producing a product at the lowest possible cost, in the shortest time, while meeting all its design requirements. The manufacturing engineer is well aware of the budgetary constraints of management in producing the product and of the desires of the customers who want the product they purchase to be as well made as described by sales and marketing personnel.

A network analyzer can be used in the laboratory to shorten design test cycles or on the production floor to reduce floor to reduce the time required for testing. (Courtesy of Hewlett-Packard)

When a decision is made to produce a product, the manufacturing engineer plans what facilities can be used and the sequencing of the production operations. If the objectives set for the product manufacture cannot be met by the equipment on hand, the manufacturing engineer specifies what type of equipment will be needed. However, manufacturing engineers get involved long before the end of the product development process.

They work with sales and marketing on an ongoing basis, adjusting the capabilities of the manufacturing plant in anticipation of new products and the time period needed for production. Manufacturing engineers also work with design engineers in determining product specifications based on current and/or projected machinery availability. They may evaluate the current overall manufacturing performance and make recommendations regarding improvements to the system. When new equipment is purchased, they assure its integration and compatibility with existing machinery. A manufacturing engineer may also be involved in the purchase of nonproducing facilities and equipment such as buildings and vehicles, judging the compatibility aspects of these acquisitions.

MARINE ENGINEERING, NAVAL ARCHITECTURE, AND OCEAN ENGINEERING

The oceans form 80 percent of the earth's surface, and transportation of goods across them and the extraction of resources from them is the domain of these engineering specialists. The naval architect designs the ship's structure; its hull form and the interaction between the hull and the water are of paramount consideration. The design must consider the control of the ship

The offshore oil drilling rig must be able to withstand hurricanes as well as the highly corrosive saltwater environment, a challenge to ocean engineers who designed it. (Courtesy of ASME)

and its stability under all conditions. Since ships are floating cities or hotels, all ventilation, water, and sanitary systems fall under the naval architect's purview, as well as the navigational systems and the propulsion system. Naval architects work with marine engineers, who are primarily responsible for the design of the ship's propulsion and auxiliary systems as well as the automatic control systems governing them.

Ocean engineers design the vehicles and devices that cannot be called a ship or a boat, such as offshore drilling rigs, offshore harbor facilities, and underwater machinery. For example, this type of machinery is used for harvesting minerals from the ocean floor.

MATERIALS AND METALLURGICAL ENGINEERING

Materials science seeks to understand the properties of materials, the fundamental bases of their behavior. What is there about the atomic and molecular structure that will make one material brittle, another elastic? Materials engineering uses this knowledge to develop new materials that will have improved characteristics—greater strength, corrosion resistance, fatigue resistance. With the new materials, new designs can be created and new manufacturing processes developed. This explains, in part, the dynamic nature of engineering, ever growing and changing. All areas of industrial production, from automobiles to hearing aids, are affected by gains in materials science.

Metallurgical engineering is the area of materials engineering that is concerned with metals. Extracting metals from naturally occurring ores—making steel from iron ore, aluminum from bauxite—is one domain of the metallurgist. Another is the development of alloys, starting with bronze in 3000 B.C.

MECHANICAL ENGINEERING

Mechanical engineers apply the principles of mechanics and energy to the design of machines and devices. Often we associate mechanical engineering design with devices that move, such as a lawnmower or food processor, but it includes thermal design as well, such as air conditioning systems. The mechanical engineer must be able to control the mechanical systems and frequently works with electrical engineers in designing these systems.

Applied mechanics is the study of motion and the effect of external forces on this motion, so the mechanical engineer is involved with engine crankshaft design and turbine rotor design. Engineers in this area must consider the vibration the device causes on the system and the counter situation of the vibrations imposed on the device. In designing the nozzle of a rocket the engineer also looks at the design from two viewpoints, the fluid's effect on the nozzle and the nozzle's effect on the fluid. There is a keen interest in the interactions between the fluid and solid interface, such as in the imparting of fluid energy to turbine blades to produce power.

Whenever there is motion, there is wear of the moving materials and eventually breakage. To minimize wear mechanical engineers must understand lubrication, choosing the lubricant that best inhibits wear between surfaces in relative motion. This is not an easy task: the lubricant must be able to withstand the operating environment of the machine and be contained within a particular space and not contaminate other regions of the device. As the mechanical engineer designs a new machine or device, he or she must be aware of new materials and their selection for gears, brakes, housings.

The field of mechanical engineering is the second largest engineering field, following that of electrical engineering.

Mechanical engineers design objects small and large. This massive crankshaft for a large diesel requires the same precision as one thousands of times smaller. (Courtesy of MAN B&W)

MINING AND GEOLOGICAL ENGINEERING

Mining and geological engineering are closely allied fields involved with the discovery and removal of metals and minerals on earth. The ore bodies may be located on the earth's surface, underground, or under the ocean floor. The geological engineer is, in general, more concerned with exploration and mapping of ore bodies than the mining engineer, who must deal with the systems necessary to extract the ore. Before a given mining method can be selected, the size of mineral deposit must be estimated by test drillings.

For surface mining, such as strip mining and quarrying, the engineer must have knowledge of soils and rocks, blasting techniques, resource management skills, and, importantly, environmental restoration. The engineer will devise plans to return the area to a natural state after the mineral deposit is removed. In underground mining the engineer must be cognizant of mining techniques, including safety considerations, ventilation requirements, draining and pumping of tunnels, surveying, and boring and blasting methods. Mining engineers are also involved with extraction of minerals from the ocean floor and below the floor.

NUCLEAR ENGINEERING

The word *nuclear* conjures up visions of disasters, from bombs to the power plant failures at Three Mile Island and Chernobyl, and to be sure the dangers of nuclear fission and fusion are real. However, the success in engineering design is overwhelmingly positive, essentially 100 percent, as operator errors have been the cause of the aforementioned calamities. Nuclear engineers also design, develop, and build systems other than power plants. For instance,

Nuclear power plants generate a significant amount of electricity in the United States (16 percent of the total) and in France (65 percent of the total) as of mid-1986. The problem of safe nuclear waste disposal is a cause for concern, and nuclear engineers are seeking ways for secure disposal. (Courtesy of Public Service Electric and Gas Company)

radiation is used extensively in medicine, particularly in combating cancer. Irradiation of food has proven to be an effective and safe method for long-term storage of bulk commodities such as grain and of highly perishable ones such as milk and bacon.

The electrical power generated by the 92 nuclear power plants in the United States was 79,000 megawatts in 1986; France produced 39,000 megawatts with 44 power plants, and Japan 24,000 megawatts with 33 power plants. These plants were designed by nuclear engineers, and the control of waste and future developments must be devised by these engineers. Materials were developed to withstand high levels of radiation. A spin-off technology from nuclear engineering is radiation techniques that detect hidden flaws in materials. The problems of fission plant design and fusion plant design and development remain for nuclear engineers to solve. One of their most significant and challenging problems is the disposal of nuclear waste and the reprocessing of radioactive materials.

PETROLEUM ENGINEERING

Petroleum and petroleum products are the lifeblood of our industrial society, and the petroleum engineer is involved with sustaining its flow. Petroleum engineers are involved in all stages of the oil and gas discovery. Working with geologists and geophysicists in locating the petroleum sites, engineers use techniques such as seismic mapping to denote the areas to be drilled. Petroleum engineers are involved in the design and use of drilling rigs, offshore platforms, and drilling equipment. These platforms are often located in hundreds of feet of water, and the drilling must extend thousands of feet beneath the ocean floor.

When a site is found, the oil or gas must not be lost to the environment and must be removed from its natural state to storage and transportation facilities. In the primary recovery phrase, the oil or gas flows under its own pressure to the surface and can be pumped. The secondary recovery phase occurs when petroleum engineers use techniques such as water flooding, gas injection, or in situ combustion to force the petroleum to the surface. Certainly the pipelines in the storage areas require the knowledge of a petroleum engineer for their design, as do the natural gas pipelines crossing the country. One of the most challenging engineering projects for petroleum engineers, working in concert with engineers from other disciplines, was the development of the Alaskan pipeline, which has to transport heated crude oil in an environmentally sound manner across hundreds of miles of Alaskan tundra.

3.5 TECHNICAL CAREERS IN ENGINEERING

Not all engineers have the same work function; the variety is quite large, depending on the type of industry or business as well as the company's size. Thus far we have looked at the types of design projects in which you might be involved within each of the specializations, but this is only one part of the panorama of opportunities that awaits you.

3.5 TECHNICAL CAREERS IN ENGINEERING **69**

A computer based signal aquisition and analysis system helps automotive manufacturers check for vibration. (Courtesy of Hewlett-Packard)

There are seven general areas of activity for engineers involved with a project, such as building a dam, or manufacturing a product, such as a refrigerator or VCR. These are research, development, design, manufacturing or construction, operations and maintenance, sales, and management. Some of the functions will be combined in companies. In addition the area of quality control and quality assurance is becoming very important in all areas of engineering, but predominantly in the design and manufacturing areas.

A new trend in many companies is to fuse these engineering areas into what is called concurrent, or simultaneous, engineering. In this approach all the areas work together from the outset to develop a given product. It has been shown in many companies that concurrent engineering reduces the time lag between product development and sales, minimizes development and product costs, while increasing product quality. All the traditional areas are involved, allowing you to specialize in one. In the course of your engineering career you will often be involved with several of these areas before selecting, in most cases, one specialization. Let's look at what is involved in each area, remembering that this is superimposed on a given field of engineering.

RESEARCH

Research in an engineering setting is not the basic research of scientists seeking to explain phenomena not understood, but rather applied research that uses the results of basic research. The development of ceramic tiles for a space shuttle or of composite materials for aircraft wings is applied research.

Challenging research in advanced engineering areas such as photonics, technologies for generating, processing, and detecting light signals. (Courtesy of Hewlett-Packard)

Why bother in the first place? In general, research is aimed at creating a new or improved product. Applied research has resulted in the creation of thermoplastics, which can be used to replace some metal parts and strengthen others in automobiles. Research creates a new technology, market conditions determine whether a new product is needed, and, presuming it is, development engineers create the product. Before we address the development function, what types of engineers are usually found in the research areas of companies? Very bright people with advanced degrees, often doctorates, who like the challenge of working in poorly defined areas of research and who exhibit high levels of creativity. Of course open-mindedness is a requirement, so that all possibilities percolate to the surface and are not discarded prematurely. Just such an attitude was required and exhibited in breakthrough developments in superconductivity.

DEVELOPMENT

Development engineers work closely with research engineers; in many organizations these functions are called R & D, because companies cannot afford undirected research. The direction must be towards product improvement or new market potential. Indeed, the development engineer must convince management to fund various new proposals based on market projections or customer requests from the sales engineer. In chemical engineering development might mean the construction of a pilot plant, in aerospace engineering, the testing of scale models in a wind tunnel. The development engineer does not always deal with the latest research, but may use existing devices in a novel way to solve a problem. When there is a problem and a new product is needed, the development engineer will create several alternative solutions.

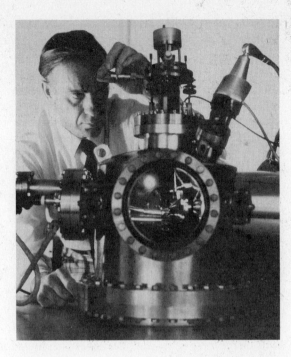

A research engineer using electrospectroscopy to detect contaminants as small as one atomic layer thick. (Courtesy of United Technologies)

DESIGN

The next step in the product or process development is taken by the design engineer. After the development engineer shows that a new product or process is possible and management is convinced to expend its resources on people, space, and money to produce it, the design engineer takes the model or concept and translates it into a producible form. The development engineer will then propose several models of a product to show that it can meet the goals established for it, such as improved efficiency or use of microprocessor control. The design engineer selects and modifies the product in light of the manufacturing facilities available: you cannot specify a tolerance smaller than the manufacturing machines can produce. The design of the product must be produced in such detail that engineers in manufacturing, though unfamiliar with the product or even its purpose, can set up the machinery, the material specifications, and the manufacturing components required to manufacture the product. The same scenario is involved for a process, such as a new paper-making technique, or for a one-of-a-kind product, such as a dam or large office building.

In designing the product the design engineer must constantly balance economics and product quality. Obviously, you want the product to be of outstanding quality, but you must bear in mind that all materials wear out with time. Consider the automobile engine: most engines will last over 100,000 miles with modest maintenance, but some components, such as filters and spark plugs, must be replaced periodically. It makes little economic sense, in light of the competition, to design an oil filter that would last 100,000 miles, as it would be very expensive. The competition's filters selling for $2 to $4 would continue to be chosen by consumers.

Not all engineering design work is done on a computer. This engineer is completing the design of a metallic filter at a drafting table. (Courtesy of Pall Corporation)

While on the topic of automobiles, a designer must consider the trade-offs involved between using normal-gauge sheet metal, thin sheet metal with thermoplastic covering for additional strength and rust resistance, or fiberglass, all the while considering the final shape of the body, its aesthetic value, and the manufacturing costs associated with each. A design engineer must have a broad engineering background and work well with the often conflicting objectives of cost, quality, and aesthetics.

MANUFACTURING AND CONSTRUCTION

Manufacturing and construction engineers face a similar challenge—to translate the paper renderings of a product or building into physical reality. This takes the coordination of many types of people, materials, and equipment. Consider that a tire company wishes to manufacure a new tire. The equipment required may be some existing machine and perhaps a new machine to perform a special function. The assembly line must be established in conjunction with existing products, the new machine ordered, materials ordered, and personnel coordinated. The manufacturing engineer must be able to communicate and work well with tradespeople, design engineers, and management. This person may have great influence on the product quality and cost, and hence the long-term viability of the company.

The same abilities the manufacturing engineer possesses are necessary for the construction engineer if a bridge or other construction is to be completed on time and on budget. The engineer must be able to handle labor disputes, quickly communicate to management significant problems that affect project completion, and assure a flow of materials to keep the work force fully utilized.

OPERATIONS AND MAINTENANCE

Once any facility exists, be it a manufacturing plant or a university, it must be maintained so that the buildings and the manufacturing machinery operate as designed. The operations engineer must be a generalist and be able to deal with electrical systems, electronics, control systems, and steam systems, and the specialists who repair and maintain them. The technicians in each area will maintain the equipment; they must be scheduled, and the schedule is affected by the analysis of the operating data from the equipment. The analysis of this data is part of the operations engineer's responsibility. Do the data indicate that the circulating water pump for the air conditioning is failing and needs to be replaced? How soon it will fail and when to assign someone to replace it without disrupting a building's work force are questions the operations engineer must answer. Computer systems require a cool environment, and businesses such as financial institutions would be in disarray if the air conditioning failed, all for lack of a pump replacement. It is with such pressure and with such awareness that the operations engineer must assess work assignments. Operations engineers are also responsible for plant safety and technician training, and must be knowledgeable about union contracts and the resulting job jurisdiction.

A construction engineer must coordinate the activities of many trades to keep a project on time and on budget. (Courtesy of American REF-FUEL)

SALES

A vital element in the survival of any company is that its product be purchased by others. Having a superior product or service is not sufficient; someone has to make it known to consumers. Sales work in engineering is different from retail sales, which is what we encounter in department stores. The sales engineer must be knowledgeable about the needs of the customer and why a product will satisfy the customer's requirements and, in all probability, will do so better than the present system or a competitor's product. The sales engineer must be a technical generalist able to deal with a variety of customers and at ease with communicating with them: this means being able to listen as well as talk. Very often it is the sales engineer who lets the company know about new product needs, and this information flows to the development engineer. The sales engineer must be technically proficient, so that she or he can relate what the advantages of a new product are to other technical people and answer their questions in technical and general terms.

Additionally, the sales engineer is the visible representative of the company, that others see and base their opinions of the company on, and thus companies are selective about whom they choose for sales. It is seldom an entry-level position because of this and because it takes time to learn the details of the product line. Usually a sales engineer starts in the manufacturing or service area, to gain a generalist background regarding the company.

MANAGEMENT

Many engineering students do not want to pursue a lifelong career in technical engineering but wish to move eventually into management. Studies have indicated that upwards of 75 percent of engineers become managers. Part of the reason for this is the societal recognition that such promotions lead to large salary increases for those at the top.

Such notables as Lester C. Thurow, dean of MIT's Sloan School of Management, have pointed out that what when it comes to inventing new technologies, the United States has no peer. Using the technologies in the industrial arena is where we lack. The managers of today are reluctant to use new technology, and by the time they adopt it, foreign competition has several years' head start. Engineers need to be involved with more than inventing the technologies; they should move into upper management to introduce technological innovations more rapidly into industry. Nontechnical managers certainly can comprehend new technology, but they do not have an intuitive understanding of which option to pursue. Engineers, because of their education and training, have developed this intuition and can learn the management skills necessary to control technical industries. It may not be sufficient in the future, as is indicated by industry's present state, to have managers in technical industries who have no technical background.

What do managers do? They manage resources—people, money, and materials. They do not create, but can be creative in the management of these resources, accomplishing a given task or implementing a phase of corporate philosophy. We all have to be managers in our daily lives, so the concept is one we are familiar with; we plan our individual and family lives and actuate the plan as best we can. We manage. However, when you are directing not just yourself but others, you need special interpersonal skills and aptitudes.

What do you do as a manager? Attend meetings, read reports and memoranda, and write reports and memoranda in support of getting things done with and through others. Thus, you must be able to plan, organize, direct, and control several tasks or activities simultaneously in support of a given objective. The objective might be to design the landing gear on a new aircraft, or to develop the specifications for a circuit board. As a manager you are responsible for and rewarded for the work that others perform.

One of the significant advantages of your enginering education and of the aptitudes you must possess to succeed in engineering is that you have many of the qualities necessary to be a good manager. Most management situations involve solving a problem, be it producing a new dishwasher or controlling vibrations in a space station. Your education is directed towards solving problems, problems in the physical world. Many of your managerial difficulties will be in solving people problems, a skill that can be learned and an art that can be refined.

GOVERNMENT, CONSULTING, AND TEACHING

Government, consulting, and teaching are the other areas where you might seek a career. Many engineers work for federal, state, or local governments in a variety of capacities. Our governments are extremely complex, exist in a technological world, and need engineers as pathfinders. Engineers provide an important link in allowing government officials to make sound decisions in light of complex technological systems. Engineers help resolve environmental questions in most public construction projects, oversee the construction and maintenance of public facilities, and direct public transportation systems and public utilities (electricity, water, and waste). In addition the government funds fundamental and applied research in the development of new technology and employs engineers to oversee or undertake such projects.

Consulting and teaching are two smaller areas of practice you might consider as career paths. Often companies find it more cost-effective to hire specialists than to keep them on staff. Thus, consulting firms originate with specialists who decide to work on their own and who have recognized expertise that they can market. You can do the same, but must develop the expertise by working in industry in a technical area. More frequently you can work directly for a consulting engineering firm that will train you as you work in their specialty areas. For some engineering fields, for example civil, architectural, and environmental, consulting in not just a "smaller area," but is a dominant area of employment.

Teaching is another area that some engineers pursue. An advanced degree, most often a Ph.D., is required, and a desire to work with students and to conduct research should be an attribute. You must be able to balance the research and teaching interests and have an independent mind to benefit from this environment.

QUALITY ASSURANCE AND QUALITY CONTROL

Quality control and quality assurance are often part of the manufacturing process, but in many organizations they are a separate department. Because of the special emphasis quality control and assurance are receiving in the United States currently, they will become a more apparent career path within organizations. With increased competitiveness in the world marketplace, any product or service must be designed and manufactured with quality control checks to assure the required performance is achieved.

Quality control means checking materials and components as they are received and manufactured, so that the steel has the correct carbon content, and the size specifications for certain materials are within tolerances. Quality assurance means checking the product or service after it is completed to assure that it meets the initial objectives. The combination of both aspects creates the quality system that is increasingly used in companies. We have all been made aware of the need to improve the product quality of many U.S. manufactured goods over the past decade, by articles in the popular press and professional journals. Companies are responding and creating stronger and more active quality assurance/control departments. They have rediscovered that quality control means greater profitability for the company.

Quality control engineers assure that the product meets the company's specifications. This engineer is checking the tire seat on the wheel rim after a test. (Courtesy of Michelin Corporation)

3.6 WHAT IS QUALITY?

Let's describe some aspects of quality, at least in regard to its use in quality systems. In the quality of a product or service, we expect a greater degree of excellence if *the price* is high compared to a competitor's, if *the cost* to

Continual testing is one element in assuring product quality. This tire is undergoing a laboratory cornering test to the point of failure. Analysis of the failed tire will yield information necessary for engineers to create design improvements. (Courtesy of Michelin Corporation)

produce it is higher, and if it is less *variable* in performance from piece to piece. The degree of excellence refers to the performance and advantages of the product prototype vis-à-vis competing products. Once the prototype is created, quality control becomes necessary, as manufactured products will vary from the prototype even under the best of circumstances. For instance, the prototype may assume a soldering technique that is difficult to perform, creating a high incidence of defects. The design should be revised in light of this information. This is the design review stage. So not only manufacturing is involved, to check that the various steps can be accomplished with the facilities available, but also quality control, to provide input regarding the repeatability and accuracy of the steps. To control the variability of the product the design must allow for the variations of perfomance of the raw materials and components. The engineers in quality control have knowledge about this variability in other situations and can help determine if it can cause problems in the product performance. For instance, an electric circuit uses a capacitor of a certain size, let's say 10 microfarads. A quality control engineer knows that the shipments of capacitors of this size will vary from 8 to 12 microfarads. The design must allow for this variation, otherwise the product may not perform satisfactorily, and the rate of returns will be too high, resulting in decreased profitability or even losses.

There are costs associated with implementing a quality control system. Appraisal cost includes the actual costs of making the quality assurance tests on the completed product and the quality control tests on the raw materials and components, as well as the cost of the design review stage. Failure costs result from a product not meeting its performance specifications. These costs arise from the remanufacture and repair of products that have been returned because of substandard quality. Accidents that occur because of substandard product performance and warranty reimbursements are both in this category. A determination made in the quality assessment of a new product, that additional training of manufacturing or shipping personnel is required, adds to preventive costs, expenses made to reduce the failure and appraisal costs.

These three categories can be combined to form a chart (Figure 3.1), depicting the total quality cost. Notice that the minimum of total quality cost assumes there will be A percent of defective products and that the quality program will incur $\$B$ of expense. To spend more effort and expense on prevention and appraisal decreases profitability, as does spending less and incurring more failure costs as a result. Inherent in this model is the maxim that product quality must be attained with limited financial resources and that the product must have a market that will return the investment. Engineers realize that companies must produce products that are competitive in price and better in performance than others on the market. This is a fundamental challenge in designing a new product. Quality begins with the creativity and

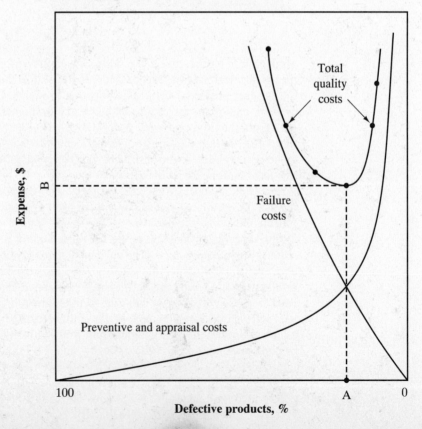

Figure 3.1 A total quality cost diagram formed by the superposition of failure costs and preventive and appraisal costs.

producibility of the design and follows through the manufacture and delivery of the product to the customer.

Developing high-quality products requires more than instituting quality-control checkpoints. It requires a corporate attitude summarized in the term "total quality management" (TQM), a concern about developing and maintaining high-quality products and services throughout the organization. Recent studies have indicated that quality is a high priority for today's consumers, corporate and individual. Ten years ago in one survey, only 30 to 40 percent of the respondents considered quality as important as price; that figure has risen to 80 percent today. Companies have also found that quality improvement and cost cutting can go hand in hand. Having products perform according to specification reduces scrap (out-of-spec parts), repair work on returns, and warranty work in the field. A large manufacturer reported these costs to be 20 percent of revenues before instituting an extensive TQM program.

There are two basic approaches to quality improvement, benchmarking and employee involvement. Benchmarking means searching for the best product or service in the industry and using that as a basis for judging your own company's performance. The concept of TQM extends beyond manufacturing a product, into distribution operations and billing as well. Once goals for improving service and products have been objectively determined, employees need to be involved in achieving the objectives.

Employee involvement can be expensive in the short term. It requires training in fundamental subjects like quality improvement, where workers learn what happens to various components when they finish with them and whether their fellow employees who use the components are satisfied with them. In addition workers conduct meetings and jointly solve problems at a level where the problems occur. Having workers solve problems at a lower level than is typical in many corporations can require management reorganization, usually to the detriment of some middle managers. Changing corporate attitudes does not come easily, a reason that giving more than lip service to TQM is difficult at times.

Companies who are willing to undertake a TQM program follow four steps. First, a variety of charts are created, plotting the performance of various sectors of the manufacturing process. The results from one of these charts tell when a machine should be adjusted, before its parts fail to meet specifications, rather than waiting for the out-of-specification condition to arrive. As a machine ages, its performance will deteriorate. By continually monitoring the output, you can have the machine serviced before the output deteriorates to the point of not meeting specifications.

Second is the design, or redesign, of products that are simpler to assemble. In general the reliability of a simply designed product will be greater than a more complex design. Third, the number of suppliers is reduced so as to encourage a closer relationship with them. By being a larger and more frequent customer your quality standards have a greater chance of being accepted. For instance, when Xerox adopted a TQM program in the manufacture of its copiers, it reduced suppliers from 5000 to 400. To maintain a good working relationship with Xerox, suppliers must maintain their expected quality standards.

Increased use of computer-controlled machinery has improved the quality of manufactured components. These engineers are discussing the output from the computer numerically controlled machines in the foreground. (Courtesy of Pall Corporation)

Last, statistical methods, based on the works of Genichi Taguchi and used in the control of manufacturing operations, are incorporated to minimize disruptions in these operations. For instance, if a process is temperature-dependent and the temperature in the plant cannot be adequately controlled, the Taguchi method would indicate that a temperature-independent process be developed. Don't continue with the existing one.

This methodology also allows for a comparison of other materials and methods of manufacture than those currently employed. Checking for improvement may be made on a continual basis.

As companies compete in the global marketplace, many must institute quality control programs that comply with international standards. ISO 9000, a series of quality assurance standards developed by the International Organization for Standardization, is essential for marketing products in Europe and assists companies such as Xerox and AT & T in the United States that have implemented compliance standards for vendors.

Concurrent engineering fits very well with TQM programs such as those fostered by ISO 9000 and allows high quality products to be produced in a more timely and responsive fashion than traditional engineering. Figure 3.2 illustrates the various functions that occur in a product development cycle with time. During the course of designing and manufacturing a product there are always changes that must be made to the design. In concurrent engineering the changes come earlier in the product development cycle when they are much less expensive. In the traditional or sequential engineering process changes tend to occur later in the cycle when they are more costly. Figure 3.3 illustrates this. For instance, a major aerospace company has found that

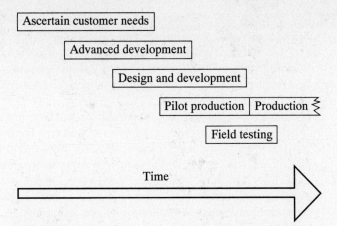

Figure 3.2 Functions in a product development cycle.

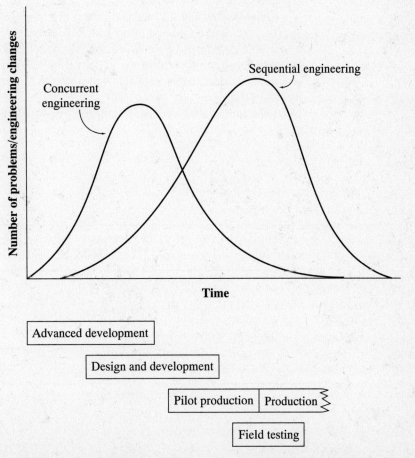

Figure 3.3 The distibution of the number of changes with time for concurrent engineering and sequential (traditional) engineering.

changes that cost $1 when made early in the development cycle could rise dramatically to $10,000 during the production phase.

Another advantage of concurrent engineering is that the time for the manufacturing cycle is reduced. Figure 3.4 qualitatively illustrates the reduction in manufacturing time. A major manufacturer of exercise equipment was able to cut its time to manufacture a new product in half by implementing concurrent engineering and total quality assurance programs.

There are changes in costs when concurrent engineering is introduced into a company. Figure 3.5 illustrates this. Notice that the spending rate is highest early in the cycle. This requires a mind shift on the part of management who often have been trained and expect that expenses will occur later in the development cycle, nearer to product manufacture and hence nearer to when revenues may be coming into the company. In addition, there are significant costs in implementing a concurrent engineering process, involving the continuous training of many people, getting people to communicate with one another, and breaking down the traditional methods of producing a product.

Consider the manufacture and design of a professional exercise bicycle. A company using TQM and concurrent engineering decided to manufacture a new model from scratch. Since the company had switched to TQM all participants were involved in the initial concept discussions. The marketing and development people indicated that the new model should be the same dimensions as the older one, in this case 46 inches high. Now that a person from shipping was involved in the initial decisionmaking, she pointed out that

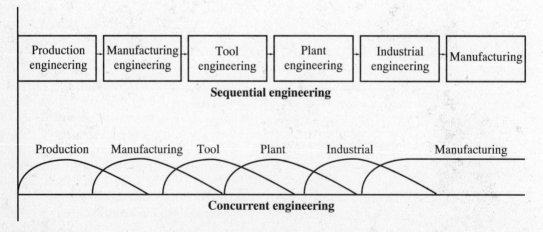

Figure 3.4 Shortening the product development cycle.

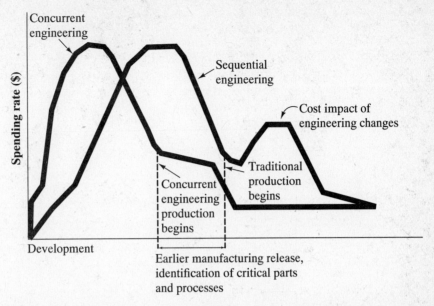

Figure 3.5 Spending profiles for concurrent and sequential engineering practice.

if the bike were 44 inches high, two could be packed on top of one another in the standard shipping container. This resulted in a savings of $50 per bicycle, which on a $1000 unit greatly enhanced the company's profit. The value of TQM and concurrent engineering did not stop there.

In further discussions involving all the personnel in the manufacturing process, a welder pointed out that the current design for the pedal bushing which holds the pedal shaft, illustrated in Figure 3.6a, required many welds. Holding the circular bushing in a precise position during the welding process was difficult and resulted in a comparatively high level of rejects. He suggested a design such as illustrated in Figure 3.6b that requires two welds and while it uses slightly more material, the rate of defects would be much lower and the quality dramatically higher than it was. This person had to feel secure that eliminating work from his functional area would not jeopardize his job. This requires continual efforts at team building and communication. There is a cost to companies for time spent on these activities, time not spent on the production line. However, increases in productivity and quality have proven in companies that the cost of continual team building is much less than the gain in increased profits and quality.

In addition the company introduced a "just in time" method of inventory and manufacture. When an order is received for 100 bicycles, the welder receives material from the supply area for these 100 units. In this situation,

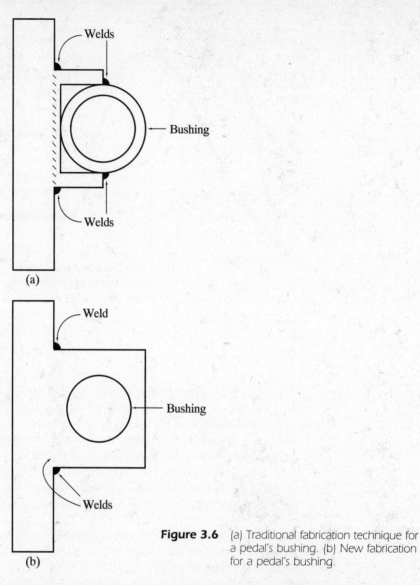

Figure 3.6 (a) Traditional fabrication technique for a pedal's bushing. (b) New fabrication for a pedal's bushing.

illustrated in Figure 3.7, the various manufacturing processes are arranged in sequence. The welder joins parts together, the machinist finishes the parts to the required tolerances, the pieces are assembled, painted and shipped. Since the manufacturing process operates as a team, if problems are noted in a certain weld, this can be corrected before all 100 units are produced. In addition, through cross training, where the welder learns the assembler's role and vice versa, the assembler can communicate which welds are visible and must look good and which ones are hidden.

This is in contrast to the traditional manufacturing practice where the organization could still be in the horseshoe shape, but the welder continually fabricates parts and puts them in a bin to be used later by the machinists and assemblers. The machinists similarly put their output in another bin and the

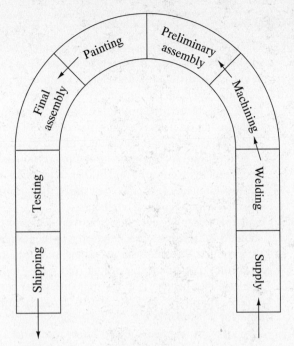

Figure 3.7 Manufacturing steps for an exercise bicycle.

assemblers and finishers do the same. An advantage to this system is that all stations are continualy busy and there is little down time. In this particular company the "just in time" manufacturing scheme had about 80% utilization, but the productivity and quality were higher. Far fewer rejects occurred, resulting in a significant net gain for the company.

"Just in time" manufacturing requires good relationships with suppliers. Imagine if the supply area could not provide the welder with material needed to start the manufacturing process because an outside vendor did not have that material in stock. The order might be lost to a competitor since it could not be filled in a reasonable time or repeat business would not occur because of unusual delays in this instance.

REFERENCES

1. Badawy, M. K. *Developing Managerial Skills in Engineers and Scientists.* Van Nostrand Reinhold, New York, 1982.
2. Beakley, G. C.; Evans, D. L.; and Keats, J. B. *Engineering: An Introduction to a Creative Profession.* 5th ed. Macmillan, New York, 1986.
3. Carruba, E. R., and Gordon, R. D. *Product Assurance Principles.* McGraw-Hill, New York, 1988.
4. Jamieson, A. *Introduction to Quality Control.* Reston, 1982.
5. Kemper, J. D. *Engineers and Their Profession.* 3rd ed. Holt, Rinehart and Winston, New York, 1982.
6. Kemper, J. D. *Introduction to the Engineering Profession.* Holt, Rinehart and Winston, New York, 1985.
7. Waterman, R. H. *The Renewal Factor.* Bantam, New York, 1987.

PROBLEMS

3.1. Investigate a recent research breakthrough and determine some of the important steps that occurred in the process.

3.2. What is the difference between development and design?

3.3. Consider each of the engineering work functions from research to sales, and discuss the aptitudes that are desirable for each.

3.4. Using a national engineering periodical of one of the engineering societies, determine the current and projected demand for engineers in general and for the type of engineer you wish to be in particular. You may also consult the Sunday classified ad sections of large metropolitan newspapers to determine the number of companies looking for various types of engineers.

3.5. Imagine a new product that you would like to produce. Then assign yourself the various roles necessary in the design and development of the product. Consider the questions and decisions you will make in each role, then determine which role seems to fit you best.

3.6. Sometimes a project is more appropriate for certain engineering disciplines than a product. Repeat the process in problem 3.5, but this time for a project.

3.7. Interview an engineer, if you know one, and assess what his or her work functions are (design, report writing, meetings, etc.) and the time devoted to each.

3.8. Explain why the demand for industrial engineers has increased faster than the demand for some other engineering specializations in the past decade.

3.9. Discuss the difference between a chemical engineer and a chemist. Interview such people if possible.

3.10. Write a report on the changing job functions of aerospace and aeronautical engineers in the past decade.

3.11. Discuss the impact of composite materials on the engineering design process in the aerospace and automobile industries.

3.12. What is a thermoplastic material? How are such materials used in the engineering design process in terms of material selection?

3.13. Discuss the impact that ceramics have on creating improved products and the material features of ceramics that make them valuable.

3.14. At the library use a reference, such as *Peterson's Guide to Engineering, Science, and Computer Jobs,* to make a list of ten large and ten medium-size companies needing engineers related to your major. Select five of the large ones and determine the high and low selling price of their stock in the past year.

3.15. For the five companies selected in problem 3.14, look for information about their financial health. Books such as *Standard and Poor's Index,* found in the financial or stock market section of a public library, give this information. What does this tell you about the company as a place to work?

3.16. Obtain an organizational chart for a company or corporate entity such as a university and examine the hierarchy. Note the variety of information that must be processed by different levels in the organization.

3.17. Major metropolitan newspapers often have a weekly science section. General interest science magazines or journals will discuss technological changes that are occurring. Make a list of the significant breakthroughs in mathematics, science, and technology that occurred in the past year. Note which breakthroughs will have impact on your field of engineering and hence on the management of companies using your field.

3.18. Examine the types of courses required for a masters degree in business administration. Can you categorize them? Contrast them with your technical courses.

3.19. Discuss why the attributes of a leader require more than technical ability alone.

3.20. What is concurrent engineering?

3.21. Describe total quality management.

3.22. Explain why costs occur earlier in the manufacturing cycle for companies using concurrent engineering than for those using traditional engineering.

3.23. Describe how the TQM approach could be applied to one of the following:
   a) design, construction and operation of a new chain of gasoline service stations;
   b) your college mathematics curriculum;
   c) design, manufacture and sale of a new line of chains for tires; and
   d) operation of your college's cafeteria.

CHAPTER 4

Ethics and Professional Responsibility

CHAPTER OBJECTIVES

To explore our values for ascertaining right and wrong.

To discover the elements of a profession.

To investigate moral dilemmas: ones you will not face and those you will.

To distinguish between product quality and product safety.

To develop a greater understanding of the pressures engineers must withstand in practice.

(Photo courtesy of the United States Navy.)

One of the cornerstones of the engineering profession is its sense of professional responsibility. An understanding of the engineer's ethical responsibility to society is implicit in being an engineer.

4.1 INTRODUCTION

It may seem odd that an engineering course should concern itself with ethics and morality; after all, we know that we should behave in an ethical fashion and have good moral values. What else is there? A lot—and this chapter introduces some of the problems and guidelines and encourages you to consider taking an introductory course in ethics.

What is ethics, and how does it relate to morality? Ethics is the application of a moral philosophy to standards of behavior. We already have opinions concerning what is right and wrong; these opinions are formed as we grow up and are based on a set of moral values we seldom question. The difficult question of what is right and wrong has been a source of discussion and analysis from the earliest human societies. When you are a practicing engineer, a contributing citizen in society, there will be times when you must face decisions that have no right answer. Realizing now that such situations occur and trying to determine the fundamental underlying issues, you will be prepared, at least intellectually, to confront morally perplexing situations in real life.

In the practice of engineering there are professional guidelines which assist in certain areas. These will be examined later in the chapter.

4.2 WHAT IS A PROFESSION?

To practice engineering you must be licensed by the state; hence you are a professional as are doctors, lawyers, and accountants. But other groups are licensed by the state, barbers and beauticians for instance, so there must be more to defining a profession than licensure. Let's examine some of the underlying characteristics of professions.

COMPLEX SUBJECT MATTER

Professionals have extensive knowledge, gained through specialized education, such that the average person cannot judge who is competent and incompetent. Society does not allow the profession to have a caveat emptor ("let the buyer beware") relationship with the consumer, as he or she cannot judge competence.

Professionals also have special skills, such as knowing where to locate precedents on legal matters, diagnostic guides in medical cases, and methodologies and equations in engineering matters. The skills flow from a body of

The manufacture of these printed circuit boards requires the manufacturing engineer to understand materials, electronics, machinery, and people. (Courtesy of Grumman Aerospace Corporation)

theoretical knowledge that enables the professional to know when to apply them.

Finally, an important aspect that distinguishes a professional from a technically competent person is an awareness of the society in which the skills are being used. Part of the educational background will be devoted to heightening cultural awareness, which is why engineering programs must have a significant portion of study devoted to the social sciences and humanities.

CERTIFICATION OF COMPETENCE

Because society depends on professions and we as lay citizens cannot judge professional competence, the state, the collective we, certifies various professions. There are at least two levels to this certification. The educational program is certified, in the case of engineering programs by ABET. Then the professional must pass a state examination, such as the professional engineer's examination. Upon satisfactory passing of the exam, the state allows you to practice in your profession and call yourself a professional.

TRUSTWORTHINESS

Since society cannot judge a professional's competence, it must rely on professionals' having an awareness of their important role in society, on their not being self-serving or taking advantage of their position. Often the profession has a self-monitoring aspect, to assure the ethical behavior of its members. To implement this policing aspect, the profession educates the member to put society and the client ahead of the self-interest of compensation. To be sure, compensation is important, but it should not be the sole driving force for the professional. The professional takes on work primarily for the psychic satisfaction of the work, and secondarily for compensation from the work as the identity of the professional merges with his or her work.

Another practical way in which professions govern themselves is by adhering to an accepted code of ethics. It serves several purposes, as a reminder of the moral standards behind the profession, as a source of guidance in difficult moral dilemmas, and as a standard for evaluating cases of alleged misconduct. The code of ethics of the National Society of Professional Engineers is reprinted in Appendix 1. Several cases that the NSPE Ethics Board has considered are included in Section 4.4.2.

PROFESSIONAL ORGANIZATIONS

Another aspect of a profession is its unique corporate identity, created by similar professionals joining together to form institutions such as ASCE, ASME, or IEEE. The purpose of these institutions is to provide service first to society and second to its members. Such organizations also help perpetuate the professional culture, developing a sense of unity among the members.

Engineering is a profession if you make it one. Herbert Hoover, the thirty-first president of the United States and an engineer, made the following comments about being an engineer.

"It is a great profession. There is the fascination of watching a figment of the imagination emerge through the aid of science to a plan on paper. Then it moves to realization in stone or metal or energy. Then it brings jobs and homes to men. Then it elevates the standards of living and adds to the comforts of life. That is the engineer's high privilege. . . . Unlike the doctor, his is not a life among the weak. Unlike the soldier, destruction is not his purpose. Unlike the lawyer, quarrels are not his daily bread. To the engineer falls the job of clothing the bare bones of science with life, comfort and hope."

The profession to which you aspire contributes much to the growth of new industry and the development of society, and should be viewed in light of its economic and political impact and importance as well as its daily function of problem solving.

4.3 MORAL DILEMMAS

A study of ethics can help us focus, define, and refine our own moral values. Most often we analyze ethical problems using our beliefs about a lot of different situations. However, our beliefs may be internally inconsistent and contradictory when we examine them. Different regions of the country and the world have different moral values. The task of philosophical reflection is to find a way to incorporate your moral judgments into a consistent and systematic whole.

The Socratic method is used to analyze positions for their logical weaknesses. Socrates questioned others as to their opinions and presuppositions and their implications, and in the process uncovered their logical weaknesses. This same method is used by trial attorneys in examination of witnesses, and can be used to discuss the three cases that follow.

The purpose of presenting the following moral dilemmas is for you to try to find the moral principal that justifies the action you would take. See if there are situations where you would use a principal in one case and not in another. What are the differences that would make you change your opinion?

As you decide what you would do and why in these situations, be aware of some pitfalls of ethical analysis. The precepts used in making moral distinctions are vague. Although we may believe it's wrong to tell a lie, most of us use "white lies," a lie nonetheless. Some percepts are presumed to be facts, when they actually have a value judgment associated with them. Often people will make a statement that appears to be informative but really is trivially true, such as "when large numbers of people are out of work, unemployment results." There is a tendency towards rationalization: when a person cheats on an examination, she may say, "Everyone does it sometimes," or "The test was unfair." If, on the other hand, someone else cheats, then she is morally indignant. This leads to another problem, what one morally *should* do and what you *would* do.

Often engineers would like to ignore such moral dilemmas and would rather deal with issues related to problem solving in the physical world. These dilemmas are problems to which there is no right solution, but they are indicative of the awareness engineers must develop if they are to participate fully in society. Being attuned to other than technical areas is part of the obligation of being a professional engineer. The political implications, the problems with no complete solution, only minimization of poor solutions, are aspects of life that engineers must constantly recognize. Engineering education tends to mask these indeterminate aspects, seeking quantifiable ones with analytical solutions.

Engineers most often work as part of a design team, making decisions about a product or service that affects the whole, but also being affected by the work of others. The overwhelming majority of engineers work for a corporation, a consulting firm, or a governmental agency and will sometimes be faced with conflicts among loyalty to their employer, to society, to their conscience, and to their profession.

CASE 1: THE OVERCROWDED LIFEBOAT

The time is 100 years ago, and a sailing ship carrying passengers and cargo to Europe hit an iceberg and sank. Fifty people survived and tried to crowd into a lifeboat intended for 20. A storm threatened, and the captain decided that some of the people would have to be cast from the boat and left to drown. The captain judged that only the strongest should remain on the boat, as they were needed to row. After days of rowing, the survivors were rescued, and the captain was put on trial for his actions. He justified his action because those who drowned would have done so anyway, and doing nothing and allowing all to perish was a worse course of action. Put yourself in the captain's situation. What would you do? Why? Is it different if you are a member of a jury judging the action?

CASE II — A PROMISE

You promised a friend never to tell anyone about a crime she committed several years ago. You have told no one, but now find that an innocent person has been charged with this crime. You plead with your friend to turn herself in to the authorities, but she refuses. What should you do?

CASE III — A LOW GRADE

You are a professor of engineering, and a student who is barely passing with a D comes to you just before the final examination and tells you that he is applying to business school for the next semester. However, to get into business school he will need a C in your course. The student seems to have been working hard but cannot grasp the material. You are tempted to give him the C, as he does not have an aptitude for engineering, and you would hate to prevent him from switching to another school. On the other hand it seems unfair to your other students. What should you do?

4.4 PRODUCT SAFETY AND PRODUCT QUALITY

Engineers are increasingly aware of product safety when designing new or improved products. Society holds the manufacturer responsible even if the product is used in a way not intended and where intermediaries have sold the product. This does not mean that a product cannot wear out, but it should do so in a safe fashion. Whereas engineers cannot foresee every circumstance in which a product may be used, they must be much more diligent in assessing how it fails. The company that the engineer works for, not the engineer, has been held to be the liable party, and her or his responsibility to society and to the company requires attentiveness in this area. A gray area exists in that product quality is involved. The company must be able to produce its products at a cost low enough to be competitive with others. To design a product that is of the highest quality and consequently has a high and uncompetitive price may mean that the company will not be able to remain profitable and be forced out of business.

The government plays an important role as it establishes the common standards for design constraints. Consider the situation in the early 1970s when pollution devices on automobiles were first used. The cars were harder to start and less economical—having poorer gas mileage—than the models of a few years before. However, the environment was improved by having fewer pollutants entering it. A company could not have initiated such a design on its own: no one would have bought the car! The same was true for the introduction of seat belts into cars. Many consumers did not like the extra cost, but once belts were mandated by the government (the collective consumer), manufacturers could include them and not risk a competitive disadvantage. One of the engineer's challenges was to design a seat belt system that is economical and meets or exceeds the government specifications.

Manufacturers cannot move too far ahead of what society expects in terms of product quality and still stay in business. With new technologies and materials available, the engineer can often redesign and improve product quality without increasing cost. In today's world marketplace engineers are competing with engineers from all nations. The need to maintain or increase product quality while not increasing, but rather decreasing, product cost, is ever more important.

Engineering has often been in the forefront of consumer protection. ASME stepped in to establish boiler standards in the late 1800s to protect citizens from boiler explosions. In 1918 the American Engineering Standards Committee was established, now known as the American National Standards Institute (ANSI). ANSI, through its working committees made up of representatives from all parties and agencies affected by a particular health or safety standard, develops these standards. The committees advise local, state, and federal agencies who in turn can adopt ANSI standards as regulations.

A marked increase in product liability legislation requires much greater diligence in the design and manufacture of products than ever before. This means that companies become responsible for products they manufactured that are misused. As a result engineers must design products that meet high standards and of improved quality, for the company to remain in business. The company may be held liable for a variety of reasons.

- If the product's design is defective, it cannot be safely used for its intended purpose.
- The manufacturing process is flawed; for instance, the testing and inspection is deficient.
- The labeling is inadequate regarding warnings and proper use of the product.
- The packaging of the product allows safety-related damage in the shipping process, or the product parts may be separated, allowing sale in a dangerous form.
- Records regarding consumer complaints, sales, and manufacturing and distribution information are not properly maintained.

Certainly this seems like a formidable list of requirements, but not an impossible list. Quality control and assurance in all phases of manufacturing from design to distribution have become more important.

ETHICAL PROBLEMS

The drive to decrease costs may lead to some situations that create ethical problems for you. For example, at the end of each month a certain number of computers, electronic components, or tools are expected to be manufactured, shipped and very importantly, billed. The bills become an asset in the receivables column of the ledger and balance the expenses incurred in producing the product in the debit column. The same event occurs quarterly as well. Businesses, especially small ones, often require short-term loans to cover the time between shipping their products and receiving payment for the products. Banks want to be assured that sufficient payments are in the pipeline, so to speak, before they will give the company a loan, to be used in part to pay your salary. As an engineer you must be aware of business needs as well as the need to maintain product standards, not letting a product be shipped to meet quota guidelines when it falls short of quality or safety standards. Quality standards are different from safety standards: the marketplace will eventually not purchase the product if it fails to meet the quality level expected of it; however, safety standards affect not just the marketplace but all of society.

There is virtually no moral dilemma in informing your superior that a problem exists with a product, when it violates the law or creates a safety hazard. Your superior will react positively or negatively to your concern. If he or she agrees, then there is no problem. The dilemma arises when the supervisor disagrees or the company decides not to change what it is doing because it believes it is following the correct course of action. If you make your complaints known outside the company, there is a high probability that you will be fired for your trouble. Perhaps the company officials are correct; you will have sacrificed a great deal, to no benefit for society, if your interpretation of what violates the law or creates a safety hazard is in error. The NSPE offers some guidelines in this area, most importantly the following: "The engineer should make every effort within the company to have the corrective action taken. If these efforts are of no avail, and after advising the company of his intentions, he should notify the client (customer) and responsible authorities of the facts." (Opinions of the Board of Ethical Review, NSPE, 1965.) In the rest of this section we will look at some situations like this engineers have encountered and how they reacted.

THE DC-10 DISASTER

On June 12, 1972, an American Airlines DC-10 nearly crashed due to a design deficiency in the rear cargo door. The door had to be secured from the outside by the baggage handlers, and people inside the plane could not check the security of the door. Should the door open, the lower cargo hold would depressurize tremendously, the passenger floor above it would collapse, breaking the hydraulic control lines for the rear engine and the tail wings, and the airplane would likely go out of control and crash.

The American Airlines DC-10 did not crash because of several very fortunate events. The captain had, by chance, practiced on a simulator how to handle the plane if he lost control of the rear engine and rear wings. Plus, the plane was lightly loaded with only 67 passengers. When the door exploded out at 12,000 feet over Ontario, depressurizing the cargo hold and causing loss of control of the rear engines and wings, Captain McCormick correctly and coolly reacted, ascertained what the problem must be, and returned the plane to Detroit.

After an investigation, the door problem was solved by installing a one-inch peephole over the locking pins. As early as 1969, however, engineers had noted the problem and suggested changes in the design. The differential pressure problem is one that has known remedies, for instance installing vents on the floor so the pressure differential could not occur. None of the solutions were implemented. On June 27, 1972, as a result of the near crash of the American Airlines DC-10, director of project engineering Daniel Applegate of Convair, the subcontractor and designer of the DC-10 fuselage, wrote a memorandum to his supervisors, stating in part, "It seems to me inevitable that, in the twenty years ahead of us, DC-10 cargo doors will come open, and I expect this to usually result in the loss of the airplane."

Nothing was done beyond minor changes such as the peephole, because of some unusual financial pressures at play. McDonnell Douglas Corporation was in a precarious financial situation and was counting on beating the competition with the newly designed DC-10; delay would allow a competitor's aircraft, the Lockheed Tri-Star, to be constructed. The assumption was that the market would not accommodate both planes. Convair, a subcontractor, was reluctant to press for a solution, since it was not clear who would have to pay for the design changes, costing millions of dollars. In response to Applegate's memo the general manager of Convair said that he was sure McDonnell Douglas would interpret the recommendation for change as an admission of error on Convair's part. The matter was dropped at that point.

On March 3, 1974, a Turkish Airlines DC-10 with 346 people on board took off from Paris. It reached an altitude of 12,000 feet, the cargo door burst open, the floor collapsed, and the plane crashed, killing all on board.

NUCLEAR REACTOR WELDS

The fate of engineers who persist in challenging management on safety issues outside of the company is not good in terms of job security, however beneficial it is to their mental health and self-esteem.

Carl Houston, a welding supervisor for Stone and Webster, a large engineering consulting firm, reported for work at the site of a nuclear power plant being built for Virginia Electric and Power Company (VEPCO) in early 1970. Having no specific assignment given to him, he was free to inspect various welding operations. He immediately found cause for concern. The steel pipes designed to carry reactor cooling water had welds that were substandard for a variety of reasons. Improper electrodes were being used at times; some electrodes did not receive required oven drying; and not all of the welders were properly qualified—indeed many were learning on the job. When Hous-

Robots perform tasks in environments that are dangerous for humans. This robot is used for routine inspection, monitoring, and surveillance tasks within areas of a nuclear power plant that are subjected to radiation. (Courtesy of Public Service Electric and Gas Company)

ton reported this to the manager, nothing was done to correct the problems, and he was told to take the matter no further. He did not take this advice and reported the matter to the head office of Stone and Webster, which again had no effect on remedying the problem. Carl Houston was forced to resign.

He then notified VEPCO and the Atomic Energy Commission (AEC) and still received no response. He notified the office of the governor of Virginia and the Virginia Department of Labor, still with no result. Finally, his two senators from Tennessee managed to convince the AEC to investigate his allegations. The AEC confirmed Houston's charges. VEPCO hired a consulting engineering firm, which also concluded that Houston was correct in finding welding deficiencies. Finally, the AEC required that the plant have three times the inspection frequencies of normal nuclear plants.

Houston was correct but out of a job. He suffered significant financial losses due to lack of employment and legal expenses.

BEING RIGHT IS EXPENSIVE

In the mid 1960s George B. Geary worked essentially as a sales engineer for U.S. Steel in the tubular steel products area. Though not an engineer by education, Geary had developed, through 14 years of working with U.S. Steel, a substantial background in the technology of manufacturing steel pipe. At this time the company was introducing a new product for the oil and gas industry, but Geary thought it had not been adequately tested and might rupture under high pressure. He so informed his superiors and suggested that additional testing be done before selling the pipe, thereby preventing financial loss due to personal injury and damage suits as well as loss of reputation for U.S. Steel. His immediate superiors insisted he proceed to market the pipe without additional testing, and Geary did so. He also went above his superior's head to corporate headquarters. Ultimately, a vice president thought enough of Geary's idea that he ordered a suspension of the sales effort until further tests were run.

On July 13, 1967, George Geary was fired. U.S. Steel even tried to prevent him from receiving unemployment compensation while he was looking for a new job, saying he was discharged for willful misconduct. Nearly a year later, the Pennsylvania Board of Review concluded that he had the welfare of the company in mind and could not be charged with willful misconduct; thus he was entitled to unemployment compensation. He tried suing U.S. Steel for wrongful discharge but lost in a closely decided case, split 4 to 3.

It seems that it takes a heroic effort to fight a company decision, even in matters of safety. This certainly could be a moral dilemma some of you will face. It is an unusual situation that will reach the dramatic instances detailed above, but not an impossible one. Engineering societies have proposed that legislation be enacted to prevent firing an engineering when his or her acts are consonant with the ethical obligation of a professional to hold paramount factors relating to public safety, health, and welfare. This legislation is making slow progress in terms of judicial trends and legislative initiatives.

Another more dramatic proposal is that if an engineer's judgment is overridden by a manager, and the engineer formally objects, arguing that this would create a serious danger to human safety or health or perhaps lead to serious financial loss, and if the manager persists in his or her decision, responsibility for the consequences rests with the manager. The manager would be legally liable if in the future such danger or loss occurred. As of now a manager has no legal responsibility for actions associated with decisions in these areas. Liability would tend to reduce the temptation to make decisions for immediate gain, such as meeting deadlines and cutting costs. Furthermore, it would make managers responsible for their actions whether or not they still worked for the company when the real effect of the decision, such as a plane crash or other disaster, occurred.

While this idea has been proposed by others, the process of drafting such legislation would be formidable; perhaps society needs initiatives in this area. It again points to the fact that engineers have to become more politically and socially aware as well as technically competent.

A POSITIVE SIDE TO PERSEVERANCE

The previous examples illustrated some of the negative forces you may confront as a practicing engineer, but it is important to note that speaking out as a concerned citizen does not usually result in dismissal actions.

In 1977 James Creswell, an inspector for the Nuclear Regulatory Commission (NRC), was assigned to inspect the start-up conditions at a new nuclear generating station, Toledo Edison's Davis-Besse facility. During the low-power testing a sudden and significant generation of heat occurred due to failure in the main feedwater system. The reactor operators did not receive clear information as to what was happening within the plant, because of faulty instrumentation and control systems. The operator assumed a valve was stuck in the open position and closed down the emergency core cooling pumps.

A similar mishap had occurred at a nuclear generating station in California, though the cause was completely different; misinterpretation of information and insufficiently clear information from the instrumentation and control systems caused operator error that could have resulted in a major catastrophe. Fortunately, none occurred at the Davis-Besse plant, for it was operating at 9-percent power and the actual valve failure was detected after 22 minutes, a relatively short time.

Creswell was disturbed that luck should be a factor in the safe operation of nuclear reactors and for over a year communicated his views to all parties concerned—the NRC, the utility, and the manufacturer of the power plant. No party was interested. He persisted. Finally, he took a day's leave and at his own expense traveled to meet two receptive NRC commissioners in Maryland. They listened and subsequently requested that NRC staff members answer the questions Creswell raised. As the memo was being typed, the Three Mile Island disaster occurred.

Unlike Davis-Besse, Three Mile Island had been operating at 96-percent power and the failure of the relief valve to close had taken over two hours to detect. This combination of operator errors created an explosive situation, in part because of deficiencies that Creswell had noted. He later received a $4000 award from the NRC. While his concern did not prevent this tragedy, his efforts and concerns for public safety were recognized. In the final analysis, his perseverance paid off.

HOLDING THE LINE

Probably everyone is familiar with the explosion of the Challenger space shuttle. The simple fact is the shuttle flew against the advice of engineers, particularly Roger Boisjoly and Arnold Thompson. They stated that the anticipated cold weather conditions in which the shuttle was scheduled to fly could create a situation where the rocket booster seals would fail. That is exactly what happened, causing a fatal explosion.

Higher management in Morton-Thiokol and NASA overruled their advice. Boisjoly and Thompson could not prove that it would happen, only that it might happen, and it was important to Morton-Thiokol and NASA to have the shuttle fly.

Following the disaster, Roger Boisjoly testified before the federal investigative commission. There was a great deal of sentiment against him, and he was able to keep his job in part because of Congressional pressure, though his status within Morton-Thiokol suffered. He offered these words of advice to engineering students at MIT the following year: "I have been asked by some if I would testify again if I knew in advance of the potential consequences to me and my career. My answer is always an immediate yes. I couldn't live with any self-respect if I tailored my actions based upon potential personal consequences."

This addresses a fundamental reason many engineers practice their profession. It contributes directly to their sense of value, their level of happiness. Aristotle noted that happiness is self-realization, not contentment. Having an easy life, according to Aristotle, is not the path to happiness; rather, using your abilities, your talents, skills, and interests, to their fullest yields the path to happiness. Continuing along this line of reasoning, it follows that the more complex and challenging the situation in which you use your abilities, the greater your happiness. Certainly the undergraduate engineering curriculum is one of the most, if not *the* most, challenging and complex academic paths available. There is a great deal of satisfaction in undertaking this challenge in school and in your pursuant career—one reason many intelligent people find engineering very rewarding.

4.5 DESIGN CHANGES

Changes are often made in the execution of a design. As an engineer in charge of a project or product, you must make sure that the changes do not affect the product or project integrity. A case in point where the changes affected the project was in the construction of the elevated walkways in the Hyatt Regency Hotel in Kansas City, Missouri. Two elevated walkways, shown in Figure 4.1, spanned the lobby. In July 1981 a dance contest was in progress and over a

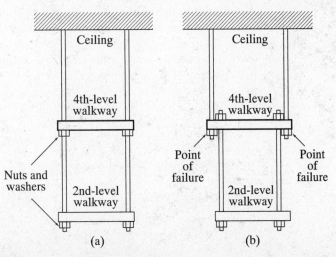

Figure 4.1 The walkway of the Hyatt Regency Hotel (a) as designed and (b) as constructed.

thousand people were on the overhead walkways, watching the participants below, with some of the observers dancing as well. The walkways were not constructed as designed but had loads twice that of the original design. The walkway supports failed, and 111 people were killed and 188 injured. Several factors led to the failure of the walkway, compounding the effect of any one factor. They included the dynamic load caused by dancing, a welded seam that had to be load-supporting, and the possible omission of some load-bearing washers.

The combination of all these events caused the failure, and the constructed walkway was of a different and poorer design than the original. Thus, not only when changes are made should all safety considerations be reviewed, but the product or project as constructed must be inspected to make sure it conforms with the design specifications.

4.6 STATE-OF-THE-ART IS NOT ALWAYS ENOUGH

Disaster may strike even when engineers have done everything correctly. Traditional methods are followed, the plans are constructed as designed, but at some point the product or project fails. Perhaps an aspect of the physical world is exposed for the first time in the new design; the fault does not rest with the engineer but with the body of knowledge that must be extended. Remember, engineers design by making assumptions about how nature behaves; when designs are pushed to their limits, the model of nature may not be sufficiently accurate and must be redefined.

A classic case in point is the Tacoma Narrows Bridge, completed on July 1, 1940; it self-destructed four months later. The bridge had a center span of 2800 feet and a width of 39 feet, making it the slenderest bridge ever built, although the fabrication and design techniques were standard. A strong crosswind blows at the narrows, and hit the bridge broadside. This caused the bridge to undulate during strong winds. In winds of only 42 miles per hour, the undulations became so severe that the bridge collapsed. Figures 4.2 and 4.3 illustrate the process. Fortunately, the problem was recognized in advance, so that people were not allowed on the bridge, and no loss of life occurred. Wind tunnel testing showed that vortices formed alternatingly on the upper and lower surfaces of the bridge, causing it to vibrate. At a wind speed of 42 miles per hour, the vibrations coincided with the natural vibrations of the structure, causing it to rip apart.

4.7 SAFETY AND RISK—RISK ASSESSMENT

Society, the collective consumer, would like all products and projects to be 100 percent risk free. That is never possible, but a design should present minimum risk to people and the environment.

Risk assessment is a combination of art and science in the goal of determining the probability of some negative event, the risk. While the calculations are quite easy (the science), the assessment as to the validity of the

4.7 SAFETY AND RISK—RISK ASSESSMENT **103**

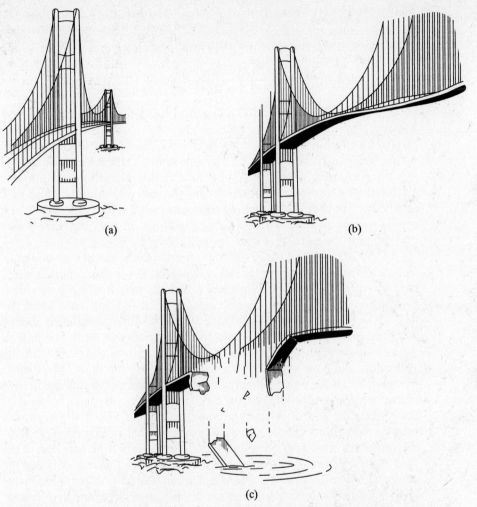

Figure 4.2 A sketch of the failure of the Tacoma Narrows Bridge: (a) the initial design, (b) the bridge undulating due to wind and vortex shedding, and (c) the bridge collapse

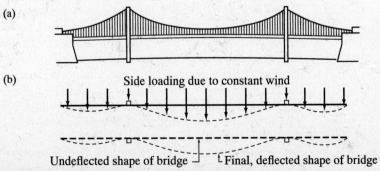

Figure 4.3 The Tacoma Narrows Bridge (a) side and (b) top views.

numbers in the calculations (the art) is not. For instance, in 1987 approximately 24,000 people died in automobile accidents with a total of 2500 billion miles driven. Thus, the risk (probability) of an accident occurring per mile is

$$\text{Probability/mile} = \frac{24,000 \text{ deaths}}{2.5 \times 10^{12}} = 9 \times 10^{-9} \text{ deaths/mile}.$$

If you were to take a trip of 100 miles, the probability that you would have a fatal accident is 9×1^{-7}, or about one chance in a million. If you were not careful in selecting the data, you might have chosen the total number of deaths, including pedestrians struck by automobiles, which for the year 1987 was approximately 49,000 resulting in a higher probability. Additional methodology may be used to further refine the data with considerations such as whether you were the driver or passenger, wore a seat belt, drove in a certain state or drove in certain weather conditions. In any case, there is data available and the calculations are straight forward.

In the area of biological sciences, the numbers are not as available, increasing the uncertainty of the risk assessment. For instance, peanut butter contains very low levels of a chemical called aflatoxin B which is produced by a fungus that attacks peanuts not carefully dried and stored and which has been shown to cause liver cancer. One study found that consuming about a 20 ounce jar of peanut butter would increase the risk of death due to liver cancer due to aflatoxin B by about one in a million (similar to driving 100 miles). But how certain are the numbers used in the study? Much more interpretation is necessary. Simply finding the number of deaths due to liver cancer per year and dividing by the total consumption of peanut butter does not provide the answer. There are many reasons for this. There are many other causes of liver cancer such as excessive alcohol consumption. In addition, there are other grains and nuts that have aflatoxin B that are part of the food chain. Further complicating the data is the fact that one does not die immediately upon consuming aflatoxin B, it can take up to 30 years and in this intervening period other events can occur which may aggravate liver cancer. A methodology used is epidemiological study where statistics are used to try and find correlations between groups of people, but it is very easy to be led to erroneous conclusions.

Biological studies rely on the use of animals and the results are extrapolated to humans as humans are not used for testing. This introduces another type of error. For instance, in one experiment to test the lethalness of a certain chemical a guinea pig required one microgram/kg of body weight, a hamster 5000 micrograms/kg of body weight, a male rat 22 micrograms and a female rat 45 micrograms. Which test animal is best to use if the results were to be extrapolated to humans? This is very important question to answer. Further complicating the situation, animal tests usually use high dose levels to determine if a chemical is toxic. The results are not necessarily linear for the animal let alone the extrapolation, creating additional uncertainty in determining the risk. Some environmental hazard testing faces the same uncertainties.

The question also arises, how does one assess the risk of new technologies where there are no negative events? The fact that a negative event has not happened does not mean that it cannot occur.

Before Chernobyl there had never been any loss of life associated with nuclear power plant accidents. Thus, there were wide ranges in what was considered the risk of danger, often based on whether one was pro- or anti-nuclear power.

The way engineers cope with this problem is to break a complex technological system down into subsystems that may have failure rates associated with them and estimate an overall failure rate. Engineers successfully estimated the risk of the space shuttle. Even though the Challenger shuttle exploded due to O-ring failure, knowledgeable engineers argued unsuccessfully to cancel the flight.

In the area of the risks due to pollution, the public may fear that not all factors are included in the risk assessment, and some may by hidden. Studies have shown that we are much more accepting of a known and uncontrollable risk (riding in a car) than a hidden or unknown risk, such as air or water pollution where not all factors can be quantified.

The perception of risk includes may factors, for instance

(a) Imposed risks seem greater than those that are voluntary, e.g., people will risk skiing, but not food preservatives.
(b) Unfairly shared risks are viewed negatively as the person receives no benefit from the risk.
(c) Risks that are personally controllable are easier to accept.
(d) Natural risks are less threatening than human-created ones. Thus naturally occurring radon found in soil is less objectionable than radon found in radioactive mine debris.
(e) The risks from catastrophes, natural (earthquakes) or human-created (Bhopal), are very frightening to people.
(f) Risks associated with advanced technologies (hormones to increase milk production in cows) are far less acceptable than those from known technologies (car and train crashes).

Table 4.1 indicates the differing views between informed lay people, a group of the League of Women Voters and a group of risk assessment experts.

If we look at areas other than hazardous materials, the matter of risk assessment is less psychologically loaded. In developing a product the applicable regulations, codes, and standards are identified, and these establish the basic safety requirements. The product concept is weighed with these on one side and the schedule, feasibility, and costs on the other side. The designer must examine safer alternatives if they exist and if a product liability lawsuit is a possibility. The engineer must consider the user in designing any product or project. This becomes very difficult for a large project, as the complexities of the systems create a staggering array of possibilities for misuse.

Disasters at Three Mile Island, a nuclear power plant; Chernobyl, a nuclear power plant; and Bhopal, a chemical manufacturing plant, were all caused by human error. A combination of events, seemingly impossible and therefore not considered in design, as a combination of safety and operating procedures were ignored, produced the disasters.

Table 4.1 RISK RANKINGS

League of Women Voters	Technology	Expert Panel
1	Nuclear power	20
2	Motor vehicles	1
3	Handguns	4
4	Smoking	2
5	Motorcycles	6
6	Alcoholic beverages	3
7	Private aviation	12
8	Police work	17
9	Pesticides	8
10	Surgery	5
11	Fire fighting	18
12	Large construction	13
13	Hunting	23
14	Spray cans	26
15	Mountain climbing	29
16	Bicycles	15
17	Commerical aviation	16
18	Electric power (non-nuclear)	9
19	Swimming	10
20	Contraceptives	11
21	Skiing	30
22	X-rays	7
23	School football	27
24	Railroads	19
25	Food preservatives	14
26	Food coloring	21
27	Power mowers	28
28	Prescription antibiotics	24
29	Home appliances	22
30	Vaccinations	25

Source: New York Times, 2/1/94, p. C10.

4.8 SITUATIONS YOU WILL FACE

You will probably face several situations involving ethical standards when you look for a job upon graduation. The first involves the recruiting process.

Suppose, as graduation nears, you send out résumés, and two companies, near each other but a distance from you, have asked you to visit them for a job interview. The companies will reimburse you for your expenses, airfare, hotel, car, and meals. Should you tell the companies that you are interviewing with both of them? Yes, though you need not specify who the other company is. You must divide the costs associated with the visit between the companies, so each pays its fair share. The cost of the interview to you is your time, but to them it's time and money. You should not charge each independently, making income for yourself on the interview process. This is the heart of the matter: not disclosure, but that you not use the visit for personal gain.

Another situation you may have to contend with is multiple job offers, but at different times. Consider the following. You have interviewed with companies A and B, and while the opportunities with both are good, you prefer those of company A. You receive a job offer from company B and have to respond in two weeks. You find the offer acceptable but hope in the two-week period to hear from company A: no such luck. Deciding to take a bird in the hand, you accept B's offer. Three weeks before you are to report, company A makes you the offer you had been hoping for. What do you do?

Accepted practice is that as long as you have not started work for the company, you may notify company B that you can no longer come to work for them and accept company A's offer. This should not be used as a bargaining tool, nor should you be under any financial obligation to company B, such as their having sent you moving expenses or having found housing for you. If you have started work for company B, it is improper to accept the offer from company A and quit work. Once you start working, company B is investing in you, training you. A certain time frame is involved before you return this investment to them, perhaps a year. Should company A still be interested in you after a year and you feel the same, then a switch is possible and ethical.

What about switching employers—how much notification should you give? Enough, but not too much; normally, between two weeks and one month. You want the company to be able to find a replacement for you. On the other hand, giving too much notice creates an awkward situation with your fellow employees. You are leaving the group and that distances you from them; your focus is where you are going and theirs is where they are. Even though employment changes are expected by the company, staying beyond a month often creates uncomfortable tension in the workplace.

What can you do now to prepare for your future career in engineering? Being an ethical and responsible professional does not start with graduation. It has to be developed now. Honesty and integrity are important attributes everyday, in completing homework assignments, writing papers, taking examinations. Trustworthiness is necessary when taking an examination, as well as when making difficult ethical decisions as an engineer. Many of the moral dilemmas we face are brought about by paying attention to shallow values, such as salary, promotion, status, power, and possessions, instead of paying attention to the more significant values of human dignity, compassion, and self-esteem.

4.9 CASE STUDIES

The following situations are drawn from articles appearing in *Engineering Times*, a publication of the NSPE, for general information to engineers. This material is reprinted with the permission of the National Society of Professional Engineers and *Engineering Times*. The cases refer to the National Society of Professional Engineers' Code of Ethics found in Appendix 1. This code of ethics and similar ones from other professional societies have several core concepts imbedded in them. The first deals with serving the public interest, protecting it from unsafe conditions; the second deals with qualities of truth, honesty and fairness, attributes we expect of a professional; and the third examines professional competence. The Code of Ethics serves as a guide to an engineer.

CASE I — WHEN IS A CONFLICT OF INTEREST NOT A CONFLICT OF INTEREST?

(*Author's note:* Before reading the case refer to the NSPE Code of Ethics in Appendix 1, particularly Sections II.4.e and III.2. These deal with conflict of interest and responsibility to society.) Here is the situation.

The law in a certain state requires every community to have a municipal engineer, whose duties and pay are fixed by municipal ordinance. Those duties vary according to the size and type of city but generally consist of attending official meetings, providing advice on engineering matters, maintaining tax maps, reviewing site plans and subdivision maps, preparing cost estimates for proposed facilities, handling complaints from citizens, and giving advice on what consultants might be needed for projects.

The city of Middleville cannot afford a full-time municipal engineer or the support staff needed for a full-time office. Middleville hires A-Z Engineers for a relatively low fee and appoints A. Z. Smith, P. E., a principal in the firm, as the municipal engineer.

Later, Smith's firm submits a proposal on a municipal contract with the city of Middleville.

WHAT DO YOU THINK? Is it ethical for Smith, as municipal engineer, and his firm to submit a proposal on a municipal contract? Smith decided he did not have a conflict of interest from his role as a part-time consultant to the city and submitted a proposal, competing with another engineering firm. Did he have an unfair advantage in making the proposal?

WHAT THE BOARD SAID. The NSPE Board of Ethical Review said it was ethical for Smith and his firm to seek the contract.

HERE'S WHY. The public interest in this case can best be served by providing small municipalities the best engineering services available. And, apparently, the state law is intended to achieve that end.

Continuity of municipal engineering services tends to ensure the best work for the city, assuming that the best available engineers are used. The engineer-consultant who becomes "municipal engineer" should, therefore, not be barred from furnishing the city with complete engineering services through his own organization if he is qualified to do the work.

CASE II — BEING PAID TO TAKE A JOB

(*Author's note:* Refer to Section II.4.c of Appendix 1 before reading this case.) Here is the situation.

Caroline Specs, P.E., a recent engineering graduate seeking employment, receives two job offers: a direct one from RZO, Inc., for a position in its sales department, and the other from Zephyr, Inc., through an employment referral firm, for a position in the design division. Specs finds the type of work at Zephyr more to her liking, but the proposed salary is $5000 a year less than that offered by RZO. After some discussions, the employment firm tells Specs that Zephyr would not increase her proposed starting salary during the first year because of its salary system for all newly hired engineers, but that the employment firm itself would pay Specs an "acceptance bonus" of $5000 if she took Zephyr's offer. She agrees to the arrangement proposed by the referral firm.

WHAT DO YOU THINK? Does this bonus arrangement mean Specs improperly received compensation from the employment agency? Was it an ethical decision?

WHAT THE BOARD SAID. The NSPE Board of Ethical Review decided that Spec's acceptance of the referral firm's proposal was ethical.

HERE'S WHY. Although the NSPE Code of Ethics states that engineers should not accept payment from either an employee or an employment agency for providing a job, the apparent intent of this provision is to prevent kickbacks from job hunters to employers. The salary offered should be the full salary without rebate or reduction.

In Spec's case, if the employment referral firm wishes to reduce the amount of its own fee by $5000, it may do so as a business decision, and that decision is not controlled by the code. There is no ethical reason to prevent Specs from choosing her preferred employment and coming out ahead financially by accepting the arrangement.

CASE III — UNFAIR COMPETITION? TO WHOM?

(*Author's note:* This case is from the *Engineering Times'* "Legal Corner," which gives capsule responses to legal questions members pose. This case has ethical overtones as well. Company X may require you to sign a form that you will not work for competing companies for a certain period if you leave the employ of company X. A major concern is that information you developed or learned at company X can be very useful to another firm, and they might hire you to obtain that information. Read Section II.1.c of Appendix 1.) Here is the situation.

I'm employed by a engineering consulting firm in Florida. Upon

CASE III Continued

entering that job several years ago, I was asked to sign an agreement stating that, in the event I left that firm, I would not compete directly with it. At that time, I didn't fully consider the implications of my signing such an agreement. Now that I'm about to leave the firm, I'm concerned about the restrictions that the agreement might impose upon me. What guidance can be provided?

The enforceability of "covenants not to compete" depends on a variety of factors, the first being whether during the period of employment you, as covenantor, were privy to proprietary information, trade secrets, or other sensitive information of the firm and, if so, whether you plan to use that information in competition with those consultants. It's not always easy to demonstrate the existence of a trade secret, and the firm may have some difficulty in defining what is confidential. If the firm has not taken steps to maintain confidentiality (e.g., it has provided liberal access to the information), your former employer may have difficulty enforcing the covenant.

Another important issue is whether the covenant is reasonable in scope, territory, and duration. For example, a covenant not to compete that prohibits former employees from competing (1) in technical areas that were not the subject of their former employment, (2) in an unreasonably large geographical area (e.g., in all of Texas, California, or the United States), or (3) for an inordinately long period of time, may be found to be unreasonable, and its enforceability may be limited.

Finally, some courts have refused to enforce "covenants not to compete" on public-policy grounds. Those courts generally hold that such provisions lack "mutuality"—employees are giving something up but they are not getting anything in return—and are inconsistent with open competition.

REFERENCES

1. Alger, P. L., Christensen, N. A., and Olmsted, S. P., *Ethical Problems in Engineering,* John Wiley and Sons, New York, 1965.
2. Grassian, Victor, *Moral Reasoning,* Prentice-Hall, Englewood Cliffs, N.J., 1981.
3. Unger, S. H., *Controlling Technology: Ethics and the Responsible Engineer,* Holt, Rinehart and Winston, New York, 1982.
4. McCuen, R. H. and Wallace, J. M., *Social Responsibilities in Engineering and Science,* Prentice-Hall, Englewood Cliffs, N.J., 1987.
5. Nef, J., Vanderkop, J. and Wiseman, H., *Ethics and Technology,* Thompson Educational Publishing, Toronto, 1989.
6. Martin, M. W. and Schinzinger, R., *Ethics in Engineering,* 2nd edition, McGraw-Hill, New York, 1989.
7. Hoover, H. H., *Memoirs of Herbert Hoover, Vol.1, Years of Adventure,* Macmillan Publishing Company, 1951.

PROBLEMS

4.1. Using the attributes of professionalism, evaluate a physician and a salaried baseball player as professionals.

4.2. Obtain a copy for the Hippocratic oath taken by medical doctors and develop a version for an engineer.

4.3. Often engineers view the term *problem solving* as a process involving a mathematical solution. Propose types of problems in engineering that are not mathematical, but require an understanding of the arts and humanities.

4.4. You are an engineer working for an oil company. The company locates a gas field near a historic trail used by settlers in the westward migration. Local citizens are upset by the prospect of drilling and transporting the gas, as the trail has historical significance and its ecology will be destroyed. Discuss the issues involved and factors that you would use in trying to resolve the conflict.

4.5. Referring to the NSPE Code of Ethics, formulate a code of ethics for student conduct in an academic environment.

4.6. The NSPE Code of Ethics suggests there are ethical responsibilities required of the engineer, the employer, the client, society, and the profession. Identify them.

4.7. It is possible for an action to be legal, but unethical. Discuss this using examples.

4.8. An engineering firm has signed a contract for the design of a small flood-control dam. The firm does not have someone on staff with this specific design experience, nor does it have the budget to hire an expert. It assigns the task to someone whose experience is nearest to that needed to design the dam. Discuss the decision in light of the NSPE Code of Ethics.

4.9. Discuss whether the NSPE Code of Ethics implies that engineering firms should provide engineers employed by the company with time off and tuition compensation for advanced degrees.

4.10. You are working for a large company and your supervisor tells you to charge your time to a project other than the one you are working on because its budget has been expended. Does this violate a section of the NSPE Code of Ethics? What are your courses of action?

4.11. Risk assessment is a major factor in value conflicts. Explain the social responsibilities engineers must exercise in developing new technologies.

4.12. List six things you value most. Try to determine how you arrived at this list, what in your background contributed to it—parents, religion, school, self-reflection.

4.13. Describe your personal ethics.

4.14. You have an opportunity to work for a defense contractor at an attractive salary with many of the benefits you want in a professional workplace. A problem with the position is that you will have to work on weapons systems. What ethical questions should you resolve, if any?

4.15. Discuss why a professional person—engineer, attorney, or physician—does not bid competitively on the performance of a service.

4.16. You and a colleague are working on a technical paper to be presented at a national conference. Before the paper is presented, your colleague is fired from the company, and your superior wants you to give no recognition of your colleague's contribution to the paper. Several alternatives are open to you: give the paper as the sole author, give the paper acknowledging your ex-colleague (risking your position with the company), do not present the paper, or take some other action. Discuss what you would do.

4.17. Jane Smith, a conscientious citizen and a principal in a consulting engineering firm, attended a town meeting where a $10 million water pollution control project was under discussion. The town was not certain it could afford the project, but would lose needed federal funds for this and other projects if it did not build the project. Smith proposed to the town that her firm would look for cost savings in the construction plan and would not charge the town anything for her services if she found no savings. The firm would charge 10 percent of the savings if it found any. In no way would it displace the design firm that drew up the plans and specifications. Was Smith's offer ethical?

4.18. A criticism of engineers is that, while they design and create devices, structures, and systems, they are directed to do so by others who, not necessarily engineers, are directing the development of engineers. Discuss the criticism pro and con, including the definition of an engineer and whether a distinction can be made between an engineer and a technician.

4.19. An automotive specialist owns a successful repair shop and has a good knowledge of complex electronic, electrical, and mechanical systems and devices. Discuss why this person is not an engineer.

4.20. Nearing graduation, a college senior decides to interview at a company in another state for a job that he has no intention of taking. He accepts the expense-paid trip. What elements of the NSPE Code of Ethics are involved?

4.21. When one rationalizes, one makes up false motives for one's actions so there are no regrets for those actions. A consulting engineer rationalized her actions for giving a kickback as follows: It was a business decision to make the kickback; no one was injured by my actions. In fact the public received a well-designed project that was completed on time. Plus, without the kickback I would have had to lay off some employees, who would then collect unemployment. Discuss her reasoning.

4.22. Using the list of topics in Table 4.1, create your ranking and compare it to that of the experts.

4.23. Discuss risk assessment as it relates to environmental hazards.

4.24. Why are biological hazards more difficult to assess than physical hazards?

CHAPTER 5

Communication— Written and Oral

CHAPTER OBJECTIVES

To explore the functions of communication.

To learn how to prepare reports and papers.

To write an effective résumé.

To understand the preparation necessary for oral presentations.

To improve your oral delivery.

To use effectively the technical information sources in your university library.

(Photo courtesy of ScottForesman.)

5.1 INTRODUCTION

Engineers must be able communicators and proficient in oral and written forms of communication.

The assumption made about most engineers is that they are very analytical, but have poor communication skills. Not only is this assumption invalid, but any successful engineer must be able to present the results of his or her investigation to others. It is important to develop your skills and abilities in this area while in college, as they will be a tremendous asset while you are in school and in your ensuing career.

Engineers must communicate more than many other professionals. For instance, suppose you are working on a project with several people in a large aerospace company. You will report at weekly meetings with your project leader on the progress that is being made. Superiors have to be informed of problems, successes, and the timeliness of the work being completed. This requires communication. It may take the form of an oral presentation at a weekly briefing meeting, or it may take the form of a written progress report. It should be apparent that the better you present yourself through clear presentations and reports, the more likely you are to be promoted. Of course your analytic skills are extremely important; good communicative ability will show them in the best light. In a survey of practicing engineers, two urged that students consider the following:

> "Communication skills, both verbal and written, are vital to any engineering career and for that matter anything you do in life. Every attempt should be made to develop and practice these skills on a daily basis, in technical and non-technical subjects."

> "Engineers should be good communicators. This is an important skill that will help them throughout their careers. A common criticism of engineers is that they tend to be too narrowly focused. By enhancing their communicating skills, engineers can expand their view and insure their views and ideas get the proper exposure. Engineers have to become sellers of ideas."

5.2 FUNCTION OF COMMUNICATION

The purpose of communication is the transmission and reception of information, of ideas and results. Communication does not exist without transmission and reception. The major error most people make in communication attempts is disregarding the receiver of the information. You must make certain that this person is in a position to understand the information as you present it. If the person is a generalist, without detailed knowledge of the analytics, do not focus on the technical aspects, but rather on the ramifications and implications of the results and the effect they will have on the project. To a technical superior you would want to communicate the analytic aspects, assumptions made and reasons for them, in modeling a certain electronic circuit or mechanical system.

The level of the language used is of vital importance. The more obtuse or grandiose it is, reflecting your command of the language and nuances of thought rather than addressing the matter at hand, the greater the likelihood that it will be misunderstood or disregarded. Engineers like straight expository writing. Be direct and to the point so there will be little cause for confusion. Engineers have an advantage over students from other disciplines; in the business world directness, even a certain bluntness, is admired. Not all academic disciplines encourage it, nor is it necessarily appropriate to the education of a philosophy major.

5.3 WRITTEN COMMUNICATION

Precision in report writing is very important, as your company most often will be subcontracting work to others or receiving contracts of work from others. This work must be communicated precisely in job or contract specifications. The specifications will detail exactly what must be accomplished, how much the company will pay or receive, and sometimes the manner in which the work must be done. Interim reports may be required. It should be apparent that being able to communicate clearly and precisely so there is no confusion for either party is vital to operating a profitable business.

In addition to the expository or informational writing of contract specifications and reports, engineers will often need to be persuasive. They need to develop their ideas and persuade others of their recommendations. This requires a great skill and careful study in writing courses.

Writing is not easy, even when you are skilled at it. Before writing the first draft, think about what you want to present, how to do it, and who the audience is. The draft is edited and finally rewritten. All successful writers, be they novelists or engineers writing technical reports, use these techniques.

As you are preparing to write a paper, a report, or a text, you must think about what it is you wish to communicate. This is the general theme of the paper. Write this down. It requires you to focus your thoughts. If it is a paper, can you tell someone in a few sentences what it is you are attempting to convey? Then move to the specifics. Jot down the ideas you wish to include as part of this general theme or topic. There need not be any order to these; let them flow. Do not break your creative thought processes. They will not form an interconnected pattern at this point, nor is it necessary that they do so. Next organize the specific points in a coherent pattern and think about how they interconnect to support the main topic. At this point you will probably have additional points you wish to make, or amplifications of existing ones. Insert these. At last you can begin to write: you have a focus, the theme or topic, and you have prepared yourself with the key points you wish to use in support of the topic. Do not be overly concerned about spelling, grammar, and paragraph structure at this point. Once a rough draft is written you can correct these defects. As you read the rough draft, correct for spelling and grammar errors. You will need a dictionary—everyone does. Use it. Nothing ruins a well-thought-out paper faster than misspellings. All writers have a dictionary at hand. A very important feature of word processing software is the "spell checker," which alerts the user to misspellings. Once you have edited your rough draft and inserted any corrections or additional comments, write your final copy. Word processing on a personal computer has a tremendous advantage because of the ease of inserting corrections without having to retype the entire paper.

5.4 THE WRITTEN REPORT

Throughout your professional career as an engineer, including your undergraduate education, you will be called upon to write reports. Most English composition classes do not cover this. A report is a researched document that relies on the incorporation of opinions of others as well as your own. Frequently, the reports you will be writing in college describe laboratory situations where experimental data must be presented, as well as the rhetoric sections.

All written reports should include certain elements:

I. The front matter
   a) title page, followed by a blank page
   b) abstract
   c) table of contents
   d) list of tables
   e) list of illustrations

II. The test
   a) introduction
   b) main body of report procedure, results, discussion, conclusion, recommendations

III. The reference matter
   a) appendixes
   b) bibliography

Not all reports will contain all these elements; however, a formal report will. Often sections of the main body are combined, such as results and discussion, or conclusion and recommendation. Since the abstract often serves as a summary, an additional summary section is not necessary. In very long reports that are intended for a general and disparate audience, an executive summary of several pages may be created, giving an overview of the report, noting salient points. However, you will not encounter reports of this nature immediately in your career.

Any report you prepare should be typed. If you do not know how to type, learn. Your grade will be a function of how good your report looks to the reader, as well as how cogently it is written. This may seem unfair, but check your own reaction to a paragraph that is handwritten on lined paper against one that is typewritten. Which would you rather read? If you have a personal computer, acquiring word processing software is well worth the expense and will improve the quality of the written reports and papers you must write during your college career. Software packages are available to teach you to touch-type, a valuable and time-saving skill.

Let's analyze what should go into a typical report that will have the aforementioned elements.

Title Page The title page includes the report title, the authors' names, the date, and the organization's name and address. For student reports the orgnaization's name is the department and college. The organization in general will be your employer; the date is important so the reader will know the timeliness of the information in the report. The information should be correctly capitalized, centered, and spaced attractively on the page.

Abstract Although the abstract is one of the first elements of the report, it is one of the last elements that is written, and often one of the most demanding. The abstract should be no more than two hundred words, often half that amount. It must contain a description of the problem being analyzed, the method of analysis, and the principal results arising from the study. Quite frequently the abstracts of technical papers and reports appear in other publications used as reference and research sources. Here is a representative abstract.

> This paper studies the rapid shearing flow of dry metal powders. To perform this study, we built and used an annular shear cell test apparatus. In this apparatus the dry metal powders are rapidly sheared by rotating one of the shear surfaces while the other shear surface remains fixed. The shear stress and normal stress on the stationary surface were measured as a function of three parameters: the shear-cell gap thickness, the shear-rate and the fractional solids content. Stresses are measured while holding both the fractional-solids content and the gap thickness at prescribed values. The results show the dependence of the normal stress and the shear stress on the shear-rate. Likewise, a significant stress dependence on both the fractional solids content and the shear-cell gap thickness was observed. Our experimental results are compared with the results of other experimental studies. (*Journal of Applied Mechanics,* 53:935)

Notice that the abstract has all the necessary elements: the problem of rapid shearing of powders, the method of solution with the experimental test facility, and the principal results concerning the normal and shear stresses.

Table of Contents As in this text, a report should have the major headings and subheadings listed in the table of contents with the appropriate page number. This allows access to those sections of immediate importance to the reader.

Lists of Tables and Illustrations The headings or titles of tables and illustrations should be given, with the appropriate page number. When there are many tables and illustrations, they are not integrated with the text but separated at the end of the report.

Introduction   The introduction is the first page of the formal report and the first topic. It should briefly acquaint the reader with the problem analyzed, what the history of the problem is, how others have tried to analyze the problem, and what you are attempting.

Procedure   The procedure section is perhaps the dullest part of any report, but critically important in that the assumptions inherent in your model used in the analysis are stated and justified. If the report concerns experimental work, you will detail how the experiment was devised and constructed and the procedure for gathering data. Computer programs will be placed in the appendix, but the methodology in developing them will be covered here. This section is essential for the reader in assessing the validity of the assumptions—you are demonstrating your technical prowess here.

Results   The results section yields that critically important piece of information—what your analysis, experiments, or survey yielded. The next section discusses these results, which is why the two are often combined in smaller reports.

Discussion   In the discussion section you give your analysis of the results. Are they as you expected them to be in light of your analysis? Why or why not? A discussion of the error involved in the results should be presented here. Remember there is an error associated with any measurement, with many computer programs, and with statistical analyses used in sampling data. You should discuss the magnitude of the errors and the effect they have on the results. In Section 7.6 we will discuss error analysis and why a scatter of data is to be expected.

Conclusion   In the results and discussion sections you made no assessment as to meaning of the results. Those sections gave the facts and an analysis of their validity. The conclusion is the most difficult section to write for many students and engineers because here you must apply your engineering know-how to explain why the results are as they are. You will be combining analytical, or theoretical considerations with the actual results to reach a meaningful judgment. In the situations you will most immediately face, laboratory reports, you will be expected to synthesize the theoretical expectations with the experimental data to explain the results. For instance, in a physics laboratory you are experimentally determining gravitational acceleration. The number you calculate from your experimental results will be different than the value given in the text. Why? In this case perhaps error analysis provides the answer. What is your conclusion regarding the experimental technique? Can your value of the acceleration be of any use?

Recommendations   In most laboratory reports recommendations are not expected, but in engineering reports this section can be the most important one to the company. Should they go ahead and manufacture a new product, will its quality level be sufficiently high, will it function well over its expected lifetime? The recommendation should reference the results and conclusions to show a logical connection and development.

Appendix The appendix and bibliography or reference section follow the tables and illustrations; either the bibliography or appendix may appear first. The appendix includes material that is too cumbersome to have in the body of the report. For instance, a set of sample calculations would be included here. There is no hard and fast rule about which material should be in the appendix, except that the report should read smoothly and the reader should not be deterred by lengthy calculations, theoretical developments, or data reduction that must be included, but whose value is in its result. This information is suitable for the appendix. For laboratory reports the appendixes would contain your calculations, the data collected, or a computer program listing. All the information necessary to perform the same analysis, for those who wish to do so, should be available in the report.

Bibliography Your bibliography lists the references you used in performing the report analysis. These will be classical text references, handbooks, and journal articles.

The following are examples of how to reference a text or handbook and a journal article.

Burghardt, M. David; Harbach, James A. *Engineering Thermodynamics*. 4th ed. HarperCollins College Publishers, New York, 1993.

Craig, K.; Buckholz, R.; and Domoto, G. "An Experimental Study of the Rapid Flow of Dry Cohesionless Metal Powders." *ASME Journal of Applied Mechanics* (1986) 53:935–42.

5.5 RÉSUMÉ OR VITA

A résumé or vita is a brief record of who you are; it will contain your biographical information with particular emphasis on education and work experience. Figure 5.1 illustrates a typical one-page résumé for a senior engineering student. It may seem unusual to include résumé preparation in a freshman course. While it is true your graduation is several years in the future, your career in engineering starts now. Résumés are increasingly important when searching for summer and part-time engineering jobs. You can plan your résumé so that in four years' time it is at least as impressive as the one in Figure 5.1. People receiving the résumé should be able to connect the job requirements with your experience. Each of us is unique, and you will have had experiences that no one else has. Your résumé should portray you as accurately and as positively as possible. Do not exaggerate, but you can discuss the qualities that show your initiative, responsibility, and engineering skills. You may want to prepare several résumé emphasizing various abilities depending on different job requirements.

Let's look at Figure 5.1. Much of the information is standard, and what personnel directors look for. If your grade point average is low, it may be best not to mention it at all. You will have to explain this in your cover letter. Your transcript will have your GPA on it, so the issue must be addressed, but not necessarily immediately. You will have to have other attributes to compensate for the low GPA. In tailoring your résumé, you may wish to change the objective slightly to better meet the expectations of a given company, if you are aware of what they want and if fits with your background.

The work experience noted in the sample résumé is not extraordinary; virtually any engineering student could have this experience if he or she

---

Jane A. Doe

Address:	194 Third Avenue New York, NY 10022 212/692-1113
Objective:	Full-time employment in robotic system development.
Education:	Top Name University B.S. in Mechanical Engineering expected May 19XX. GPA 3.2/4.0
Honors and Awards:	Dean's list: 4 terms Tau Beta Pi, National Engineering Honor Soc.
Experience:	Top Name University, Advisement Office: September 19XX – present; tutor in sophomore engineering courses for international students. Marshall Engineering, Inc.: Summer 19XX; assisted engineers in laboratory testing of pneumatic and electronic control system components; wrote preliminary sections of final report. Fastfood Chain, Inc.: Summers 19XX and 19XX; worked in a variety of positions; promoted to assistant shift supervisor, responsible for work of 8 others.
Activities:	Vice President — TN University Chapter of Society of Women Engineers, 19XX–XX Member, American Society of Mechanical Engineers Senator, TN University Student Senate Intramural sports: volleyball, handball Hobbies: hiking, photography, tennis
Personal:	Citizenship: United States of America
References:	Dr. Richard Jones Mechanical Engineering Department Top Name University New York, NY 10001 Dr. Sandra Degas Mechanical Engineering Department Top Name University New York, NY 10001 Mr. James Earl Senior Engineer Marshall Engineering, Inc. 1025 East Road Flemington, NJ 15231

Figure 5.1 An example of a résumé for an engineering student.

wished. Not everyone can work for three summers in engineering firms. Many of you will work for stores and other commercial enterprises. Use them to your advantage: if you were promoted, bring it to everyone's attention. This means someone else thought well enough of you to promote you—you are a valuable person. The other employment situations show that you work well with others and have exercised initiative and responsibility.

You will be writing a letter to accompany the résumé. It provides you with the opportunity to briefly amplify certain aspects of your educational and work experiences. Figure 5.2 shows a cover letter for the résumé in

---

194 Third Avenue
New York, NY 10022
15 April 19XX

Mr. Charles Foster
Director of Personnel
Precision Engineering Co., Inc.
Cambridge, Massachusetts 12062

Dear Mr. Foster:

I am very much interested in working for Precision Engineering Company in the area of robotic systems and have enclosed a résumé of my related work experience.

I would like to amplify briefly certain sections of the résumé to give you a clearer picture of my abilities. Through my work as a tutor, I have gained a true understanding of engineering fundamentals and have found that not only was the opportunity to help others rewarding, but my comprehension of complex subject matter is more complete.

While employed by Marshall Engineering I realized that a career in engineering was indeed the right choice for me. I enjoyed the commercial realities that engineers must confront and worked well with the senior engineers. I feel very proud that they trusted me enough to allow me to prepare the introduction and test sections of the evaluation reports on the G26 and E28 control systems.

I believe that my background shows that I contribute to every organization I am affiliated with, be it student government where I learned the give and take necessary for group decision making, or Fastfood Chain, where I was promoted to assistant shift supervisor at the age of 19. I know that I can contribute to your company and look forward to hearing from you.

Sincerely,

Jane A. Doe

Figure 5.2 An example of a covering letter for an engineering student's résumé.

Figure 5.1. Remember the purpose of the résumé and the cover letter is not to get you a job; you will have to do that in person. It is to interest your prospective employer in you, to show that you will be the type of employee that his or her company is looking for. During the interview be prepared to expand more upon your résumé. The interviewer will use this as a basis for asking questions and developing a conversation with you to assess who you are as a person.

5.6 MEMOS

Much of the writing that you will perform and receive as an engineer will be in the form of a memo. You will frequently receive directions to proceed on a project by memo, and you will query others in the organization via a memo.

A memo should be brief and to the point and include the following information:

Why you are writing the person;

What your purpose is;

What you want the person to do;

When you want it done.

Figure 5.3 illustrates the typical memo format—to, from, subject, date. This memo also indicates a preliminary design project that an engineer is to initiate often encountered in engineering practice. In Chapter 9, we will investigate how to deal with this type of design memo. Note that it is very open-ended and will require some memos from the staff engineer to gather more information.

A memo should be brief and on target, not addressing more than one topic. A common error that people display in memo writing is a lack of focus on the problem and what is expected of the respondent.

5.7 ORAL COMMUNICATION

You will frequently need to make oral presentations to colleagues at weekly meetings or conferences. It is quite normal at first to be a little anxious, and you can take steps now that will help you throughout your career. One of the first is to take a course in speech, perhaps as part of your humanities requirements. In some engineering classes you may have reports to present orally; be sure to participate. Few of us relish the opportunity to do this, but the experience builds poise and confidence. If you cannot give presentations in these areas, look for less threatening areas for initial participation, such as in church or civic organizations. For most of us the nervousness disappears after the first minute or two, especially if we are well prepared and we become involved with the topic under discussion. Few of us can stand up and talk about a subject without preparing what it is we want to say. Just as with writing, you

VILLAGE OF KENSINGTON

MEMORANDUM

TO: Frank Smith, Staff Engineer
FROM: Joan Kenyon, Chief Engineer
SUBJ: Traffic Light Installation at the Corner of Oak and Main
DATE: 9 September 1994

Traffic congestion at the intersection of Oak and Main has become increasingly dangerous with several accidents occurring just last week. The mayor wants us to replace the stop signs with traffic lights. You'll need to position the lights far away from the curb line as there are trees near the curb. Also, it's easier to see traffic lights in the middle of the street. There are several steel street light poles in storage at the Public Works garage which you may be able to adapt or use as is.

We need to start work as soon as possible, before winter arrives. Please send a memo regarding possible design options and your recommendations by the beginning of October.

Figure 5.3 An example of a memo.

must have a clear idea of what it is you want to express. Initially jot down the facts and the integrating theme that connects them, and build your presentation, oral or written, with this in mind.

Normally you will know who your audience is and what level of presentation is required, and your presentation should be prepared to suit their background. Your talk should not be too elementary nor too sophisticated, as in either case you will lose contact with the audience. Remember communication is a two-way street—transmission and reception.

Engineers seldom work in isolation and need to be able to clearly communicate with their colleagues. (Courtesy of Michelin Tire Corporation)

In most circumstances you will have 15 to 20 minutes for your presentation. If it is any longer, the audience loses interest. It should be apparent from your presentation that you are secure and in control of the subject matter. You will be asked questions, and this is often the most informative and important part of a presentation. You must be prepared to expand on given topics.

Preparation How should you give the talk? Use notes, do not read. Beforehand, prepare an outline of the key topics to be covered. Expand on these topics in your presentation. You should have gone over what you want to say many times before actually makig the presentation. Once you believe you have all the information needed to make the presentation, prepare note cards from the outline and give a trial talk to yourself. Time it. Remember to speak clearly and not too fast. Once you are satisfied that all the information you wish to convey is in the material you have prepared, try the presentation out on a friend or co-worker. As you are preparing the talk, do not use jokes to make points or to increase the receptivity of the audience. In general they will fail and should not even be attempted.

Visual Aids Visual aids are a terrific help. They provide something for the audience to read while listening to you, and help you maintain the awareness of the point under discussion. The three methods used in most circumstances are paper charts, often flip charts that are about two feet by three feet; $8\frac{1}{2}$-by-11-inch transparencies; and 35-mm slides. Imagine yourself in the audience: which visual aid is required? Flip charts work well in small groups, 35-mm slides for large groups; for other groups transparencies are a good choice. If you are using transparencies or slides, be sure to view them before the presentation, to be sure that they look the way you want them, that lines are not blurred, for instance. If possible the projection distance for the preview should be the same as for the event.

A second great advantage to using visual aids is that they are your outline and notes. You can expand upon the topics listed on the chart or graph, and often do not have to use notes at all. When constructing the diagrams and charts, keep them simple. Remember that your audience will be trying to listen and read at the same time. The visual aids should augment your presentation, but complicated diagrams are sure to turn off your audience. The slide should contain key words, graphs, diagrams, or photographs that enhance what you have to say. Do not read the slides to the audience: if you have nothing to add to what is on the slide, they do not need you—they can read.

Presentation Techniques

Just as the appearance of your written report affects how people view what is within it, so your appearance affects audience reaction to what you have to say. Not only should you look presentable, neat, and well dressed, but you should behave properly. Speak clearly and loudly enough so the audience can hear you; people do not want to strain to understand what you are saying. And speak slowly enough so they can follow. You know what you want to say and what the connections are, but they do not, and you must allow time for these connections to occur.

Talking quickly and softly are signs of nervousness you can be aware of and correct. Avoid the "uhs," the "likes," and the "you knows." Often you will have a podium to hold your notes and to hang onto initially. Stand erect and do not put your hands in your pockets. Some men will thrust a hand in a front pants pocket throughout their presentation, often jingling change. This is a terrible habit, and totally distracts the audience from what is being said.

Important to the audience and to you is maintaining eye contact throughout your talk. Look at people directly, not the same person constantly, but shift your attention from one to another. This keeps you involved with the audience, makes you aware that your message is getting through, and lets the audience know that you are engaging some of its members constantly.

5.8 LIBRARY USAGE

In your career as a student and as a professional, being able to access information about a particular problem is very important. The library is your best source of information in all areas. Being able to use it effectively will enhance your abilities. You will not need to spend undue time trying to locate a piece of information, and you will be able to locate the information that will help solve your problem. There are no courses available to you in library usage, though often the library staff will show you the resources available. Using the library resources wisely will help you immediately in writing term papers, laboratory reports, and senior papers. Following graduation you will have an ability that few of your colleagues will possess. Locating information necessary for a project, rather than reinventing it, is a valuable skill. This skill also prevents failure altogether, if the information is too difficult to reinvent.

An engineer usually relies on one, perhaps two, indexes and a few major journals that specialize in her or his area of interest. Journals are compilations of technical papers that professionals have written in a technical field, such as heat transfer. Various organizations, including professional societies, publish these journals to help disseminate the technical progress that is being made. An index is a periodic publication, monthly or quarterly, that lists by topic the articles that have appeared recently in a variety of journals.

You may be surprised to learn that when you are looking for information on a given topic in research journals it will not be located in one information source. Instead, approximately one-third of the articles of interest appear in primary journals, one-third in related journals, and one-third in unrelated journals. For example, one may find articles on heat transfer in the *Journal of Heat Transfer* (primary journal), the *Journal of Applied Mathematics* (related journal), and the *Journal of Food Science* (unrelated journal). All of these journals could have articles related to the same problem, and without proper searching, repetition of work and delays may result.

Data Base Searches Data base searching of information is provided by a number of services that have information similar to that in the indexes, discussed below, but that have greater searching abilities. These data base services have two major advantages: they are faster than searching printed documents, and they are more current, as the publishing cycle is shorter—the information is available once it is put into the computer data base. The Venn diagram in Figure 5.4 shows how these services operate. Let's look at a journal article titled "Stresses in Submarine Superstructures." Each circle, or set of information, will contain the key word *stress, submarine,* or *superstructures*. The intersection of all three sets will yield those articles that have information about all three key words. A computer can determine this quickly and print out the article information for you. For most students the data base services are not readily available, and you will be using published indexes as your current reference source. In addition, more libraries have information stored on CD-ROM programs for similar retrieval.

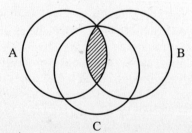

Figure 5.4 A Venn diagram illustrating on-line searching of a database.

Scientific Encyclopedias and Technical Handbooks

Let us assume that you want to investigate an area that is new to you. The first place to look would be in an encyclopedia such as the McGraw-Hill *Encyclopedia of Science and Technology;* there are others relating to chemical engineering, biology, and so on. The encyclopedia has an overview of the particular subject by an authority in the field and includes references. In conjunction with the encyclopedia, the various handbooks in particular field may be used (e.g., *Mechanical Engineers' Handbook, Electrical Engineers' Handbook.*) They contain a wealth of information in compact form. Often references are cited that are recognized as basic source material. Using these references is the next step; they help you define and better understand the problem before proceeding to current literature.

At this juncture two paths are open. You can consult the card catalog under the particular subject heading or start a search of the current literature.

Card Catalog

Most physical card catalogs have been replaced by a computerized version. However, the computerized versions will have subject headings for books in various fields, such as diesel engines, themodynamics, and electrical circuits. Included on the subject card is a brief content summary of the book. Figure 5.5 shows a typical listing. The title, author, publisher, copyright date (1986), and total number of pages are listed. The text is part of a series published in mechanical engineering by Harper and Row. Most importantly, the call number tells you where to locate the book.

BURGHARDT, M. DAVID
   Engineering Thermodynamics with Applications. 3d ed.
New York; Harper & Row, © 1986, 608 p., ill.; 24 cm
(07945539). The Harper & Row series in mechanical
engineering.
Includes index.
Bibliography: p. 516
1. Thermodynamics I. Title II. Series
CALL NO.: TJ 265.B87 1986 copy 1

Figure 5.5 A card catalog listing.

Indexes

To find current works, you will use the various indexes at your disposal. Perhaps the most familiar to most scientists and engineers is the *Index of Applied Science and Technology*. This monthly index has various subject headings under which articles from 335 journals are classified. This index has no author listing; hence, there is no ready method of checking whether a particular author has published anything or not. Figure 5.6 shows a typical listing from the *Index of Applied Science and Technology*. The article title is the key to indexing the article, and you may be misled or miss relevant articles because of their title; this highlights the importance of a title when writing a report or paper.

Heading: **Laser beam welding**

Modelling the fluid flow in laser beam welding. M. Davis and others. bibl il diags *Weld J* 65:167s-74s Jl '86

Explanation: An article on subject of laser beam welding entitled, "Modelling the fluid flow in laser beam welding," by M. Davis and others, with a bibliography, illustrations and diagrams, will be found in *Welding Journal*, volume 65, pages 167–74 in the July 1986 issue.

Figure 5.6 A sample entry from the *Index of Applied Science and Technology*.

The *Engineering Index* is a second popular index, which selects articles of engineering significance from approximately 2700 journals and periodic publications. It includes an author index, so that if you want to know whether an author who publishes in a certain field has published anything recently, you can check directly. A further asset of this index is that a brief abstract of the journal article is given along with the title. Figure 5.7 shows an entry from the *Engineering Index* for the same article as in Figure 5.6. Notice the increase in information that you have available to you. The author's abstract has been edited in this case, but you have a clearer idea of what is presented. Furthermore, you have information to allow you to contact the author if you are pursuing research in a similar area. A limitation to both indexes is a delay

Heading → Welding, Electric—See also Metal Forming—Laser Applications, Polyethylenes—Welding.

Subheading → Laser—See Also AUTOMOBILE MANUFACTURE—Welding; CONTAINERS—Steel; HEAT TRANSFER—Disks; MAGNESIUM AND ALLOYS—Welding; METAL CUTTING—Laser Beam; NUCLEAR FUELS—Cladding; SAWS—Diamond; WELDED STEEL STRUCTURES—Fatigue; WELDING—Dissimilar Metals.

EI abstract number → 128398 MODELLING THE FLUID FLOW IN LASER BEAM WELDING. ← Title of article The use of lasers in welding is becoming increasingly important in high technology industries. After a brief review of the literature, analytical and numerical results of two-dimensional models of the flow of molten metal around the keyhole are discussed. This work illustrates the connection between the absorbed ← Abstract laser beam power, weld width, weld speed and size of the keyhole for different metals. In conclusion, some insight is gained into fluid dynamical flows of molten metals in the process. (Edited author abstract) 11 refs. ← Number of references

Author's name and affiliation → Davis, M. (Univ of Essex, Colchester, Engl); Kapadia, P.; Dowden, J. *Weld J (Miami Fla)* v 65 n 7 Jul 1986 p 167s–174s. ← Abbreviated title, volume issue, date, and pages of source publication

Figure 5.7 A sample entry from the *Engineering Index*.

of up to a year before the journal article is indexed. It takes time to receive the journals, decide on their heading and the location in the index, input the information, print the index, and distribute the index to the libraries. This is a continual process, and six months' lag time is common.

Delay is inherent in any index, but the *Science Citation Index* overcomes that to a great extent. This index covers approximately 3800 journals and periodicals. The index may be broken into several distinct parts: citations, sources, and subjects. The citation section lists prior articles, by author, that have been cited in current periodicals. It includes the current articles citing the prior one and the total number of citations for a particular article. This gives a key to the relative merit of the cited article. An often-cited article will probably be pertinent and authoritative. Einstein's work is still being applied and thus is often cited. The source section is a listing by author of current articles that have been included in the *Science Citation Index*. The subject section is unique. The key words in the article are permuted. Thus, a journal article such as "Stresses in Submarine Superstructures" would have under the heading "Stress" the key words, *submarine* and *superstructures*. After the key word, the author's name is given and can be looked up in the source section to find the journal article. The words *submarine* and *superstructures* are also presented as subject headings, with the remaining key words in the title listed below it. The *Science Citation Index* takes some practice to use but is more timely because minimal editing is required before the information is published and distributed.

Government Documents
In this day and age of government-sponsored research, be it funding of a private company or of directly supported research laboratories, no search could be complete without an investigation of the many government reports. Fortunately, the *Government Reports Index,* which indexes all the technical reports monthly, is available. The several different types of indexes include subject, author, and report number. The use of government reports should not be underestimated, as much of basic and applied research is carried out with government funds.

REFERENCES

1. *The Chicago Manual of Style.* 13th ed. University of Chicago Press, Chicago, 1982.
2. Strunk, W., Jr., and White, E. B. *The Elements of Style.* 3d ed. Macmillan, New York, 1979.
3. Turabian, K. L. *A Manual for Writers of Term Papers, Theses, and Dissertations.* University of Chicago Press, Chicago, 1967.

PROBLEMS

5.1. Prepare your own résumé using Figure 5.1 as a guide. Think about what you can do during your college career to enhance your qualifications.

5.2. Look at a simple piece of equipment with no moving parts, such as a window frame, a curtain rod, a door. Write a specification so another person could construct this device. Share your specification with a classmate and see if he or she understands exactly what to do.

5.3. Go to the library and find the encyclopedias, handbooks, card catalog, and indexes mentioned in the text. Locate engineering texts on electrical circuits, electronics, thermodynamics, strength of materials, and aerospace engineering.

5.4. Consider the following general topic areas: robotic arm design, biomechanics and prosthetic devices, advanced composite materials for aerospace application, rocket propulsion systems, intergrated circuit design, microvolt measurement in humans, fiber optic transmission, bridge design, and automotive engine design. By consulting encyclopedias and handbooks, narrow the area to coincide with your interest, and write a one-paragraph design specification of this area. Then consult the card catalog and list any texts, not more than three, that pertain. List five current references from the scientific and engineering indexes available to you.

5.5. Prepare a five-minute talk on one of the areas mentioned in problem 5.4. Develop flip charts and notes, and listen to yourself give a mock presentation. If possible do this with several classmates, and critique each other.

5.6. Select a journal article that you understand, and write an abstract for it. Does it correspond to the one prepared by the article's author? Why or why not?

5.7. Critique your essay from problem 1.1 in light of the information in this chapter. Rewrite it.

5.8. Imagine that you would like a sidewalk constructed between two points on campus. Write a memo to the plant engineer indicating this.

5.9. Imagine that you are president of an engineering club and your organization will be holding a conference on campus. Write a memo to the engineering department chair seeking assistance on specific topics. (You decide what topics will require assistance.)

CHAPTER 6

Graphical Communication

CHAPTER OBJECTIVES

To learn the elements of graphical communication that complement verbal communication skills.

To become familiar with presentation techniques for qualitative and quantitative data.

To practice the fundamentals of plotting on semilog and log-log paper.

To use sketching fundamentals.

(Art courtesy of Biblioteca Ambrosiana, Milano, Italy.)

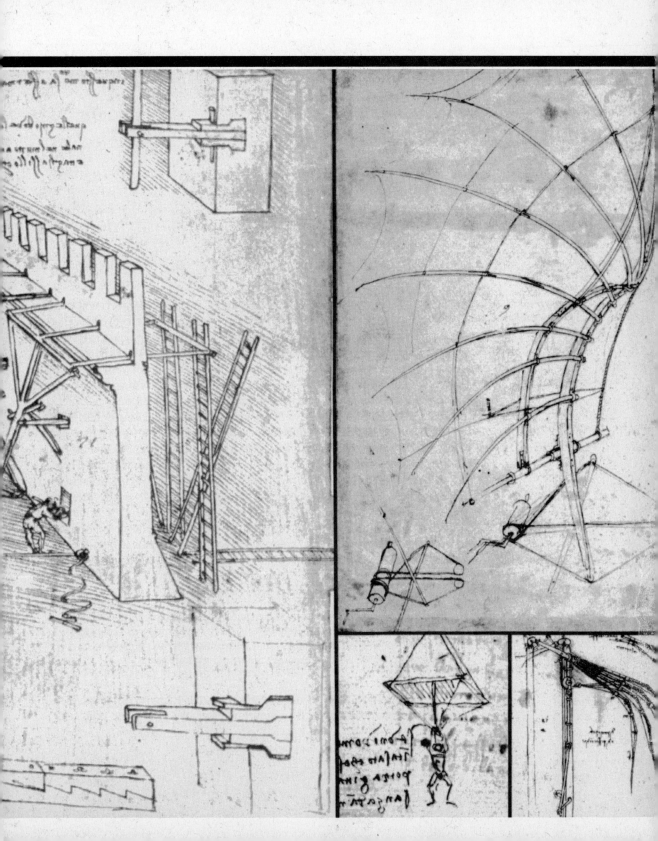

In this chapter you will complement your communication skills with the introduction of elements of graphical communication.

6.1 INTRODUCTION

The adage "a picture is worth a thousand words" is true in engineering. Often a graph can be used to display a trend or relationship between variables that is not apparent from looking at numerical data. Hence, graphical analysis is an important engineering skill, augmenting written and oral communication skills. Additionally, you will have to make sketches as part of your everyday engineering life. These sketches help you understand a problem you are trying to solve, such as the forces on a certain moving part, and communicate ideas to others. Very often you will want to sketch an object to complement your verbal description. As an engineer you will be responsible for reading and approving drawings made by a skilled draftsman, and although you do not need the skill to produce such drawings, you must be able to read them. These drawings are still called blueprints, referring to the printmaking process in which the lines were white on a blue background. Now, however, the lines are usually blue and the background white. Figure 6.1 illustrates a blueprint of a shipboard refrigeration system. The output plotters used in computerized drafting allow a variety of line colors, but the term *blueprint* will still mean a line drawing, regardless of the color. A practicing electrical engineer commented in a survey:

> "An important observation regarding engineering education is the knowledge of drafting and interpreting design drawings. The 'blueprint' is the language by which engineers communicate, and I have found far too often that my peers lacked basic skills to generate and interpret blueprints."

In graphical communication you must consider the type of information that you are presenting as well as your audience. We can divide graphical

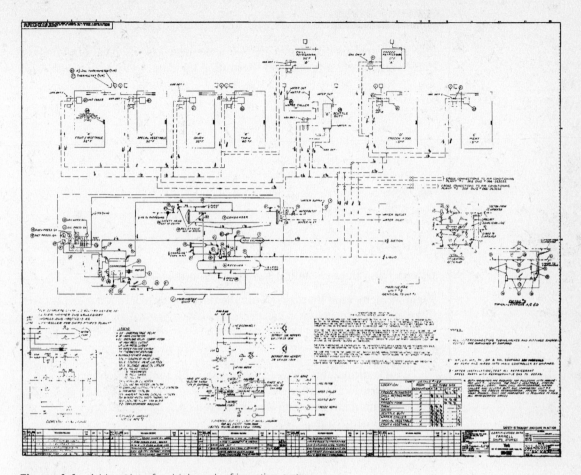

Figure 6.1 A blueprint of a shipboard refrigeration system.

analysis into two general types, one where the information is qualitative in nature, using such techniques as pictographs, bar graphs, and pie charts, and the other where the information is quantitative, using line graphs.

6.2 PRESENTATION OF QUALITATIVE RESULTS

In making a presentation you may want to give people a general understanding of the magnitude of the terms, without presenting the specific numerical data. In such a case pictographs are useful.

EXAMPLE 6.1

The following data represent the world natural gas reserves in trillions of cubic feet. Create a pictograph illustrating the data.

Africa	209
Asia Pacific	120
Europe	142
Middle East	513
North America	288
Russia	918
Other	114
Total	2304

SOLUTION

Given: The values of natural gas reserves in the world.

Find: A pictograph that represents the data.

Assumptions: None.

Analysis:

1. Determine the size of the fundamental unit for each picture and what to use as the picture.
2. Determine the number of units required for each item.
3. Construct the pictograph.
4. Label and title the graph.

You need to have a picture such that the smallest quantity has some part of the shape and the largest quantity does not have too many items. In this case the largest value is 918 and the smallest is 114, so let 114 be the quantity of the smallest picture. The picture in this case is an oil and gas derrick. When we divide the total production by 114, the following figures are obtained. Note that it is impossible to draw small fractional parts, and they have to be rounded off. This is in part what makes this presentation technique qualitative.

Africa	209/114 = 1.83
Asia Pacific	120/114 = 1.05
Europe	142/114 = 1.25
Middle East	513/114 = 4.50
North America	288/114 = 2.53
Russia	918/114 = 8.05
Other	114/114 = 1.00

The pictograph is:

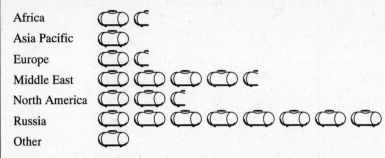

Figure 6.2 A pictograph of natural gas reserves by geographic region. Each symbol represents 114 trillion cubic feet of natural gas.

A circle or pie chart illustrates the fraction or percentage of the whole for each component. The circle chart illustrates the size of each component relative to one another as well as to the entirety.

EXAMPLE 6.2

Construct a circle chart from the data in Example 6.1, where the circle represents the total world gas reserves.

SOLUTION

Given: The values from various regions for natural gas reserves.

Find: A circle chart to represent this data.

Data: Refer to Example 6.1.

Assumptions: None.

Analysis: Determine the fraction or percentage of the whole for each component and then convert this into degrees. The fractions should add up to one and the total degrees to 360. It may be necessary to adjust slightly one or two values so the total is correct.

Africa	209/2304 = 0.0907	0.0907 × 360° = 33°
Asia Pacific	120/2304 = 0.0521	0.0521 × 360° = 19°
Europe	142/2304 = 0.0616	0.0616 × 360° = 22°
Middle East	513/2304 = 0.2227	0.2227 × 360° = 80°
North America	288/2304 = 0.1250	0.1250 × 360° = 45°
Russia	918/2304 = 0.3984	0.3984 × 360° = 143°
Other	114/2304 = 0.0495	0.0495 × 360° = 18°
Total	= 1.0000	= 360°

The circle chart, drawn and labeled with this information, is shown in Figure 6.3. If the data is put in a spreadsheet, Figure 6.4 results.

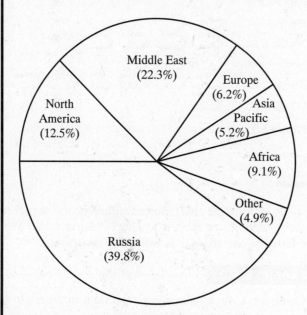

Figure 6.3 A circle chart of the world natural gas reserves by geographic region.

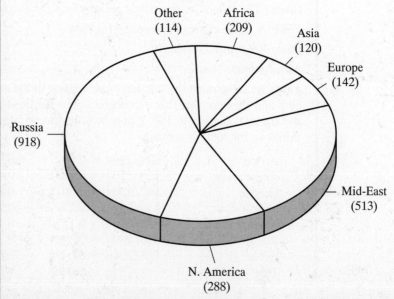

Figure 6.4 A spreadsheet of the world natural gas reserves by geographic region; produced from data in Figure 6.3.

The last type of qualitative chart we will address is the bar chart. This chart can be used in the same manner as the pictograph, or can be used to show the variation of quantity over periods of time.

EXAMPLE 6.3

Construct a bar chart to depict the average monthly rainfall in inches from the following data.

Month	J	F	M	A	M	J	J	A	S	O	N	D
	2.5	1.9	2.9	3.1	3.4	2.0	3.9	4.6	3.9	2.7	2.1	2.8

SOLUTION

Given: The data for the average monthly rainfall in a know locale.

Find: The bar chart that depicts the data.

Assumptions: None.

Analysis: Decide whether vertical or horizontal bars will look better. Determine the bar length that allows the maximum amount to be nearly full scale. Next determine the total number of bars, and locate them to cover the space available. Finally, construct the bars, and label and title the chart. The vertical bar chart for the rainfall data is shown in Figure 6.5. A spreadsheet plot of the data would look like Figure 6.6.

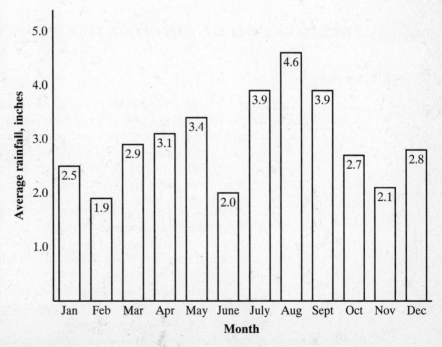

Figure 6.5 A vertical bar chart of average monthly rainfall.

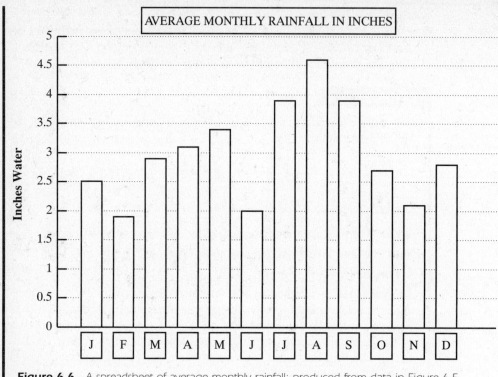

Figure 6.6 A spreadsheet of average monthly rainfall; produced from data in Figure 6.5.

6.3 PRESENTATION OF QUANTITATIVE RESULTS

LINE GRAPHS

Line graphs are the most commonly used form for displaying graphical information; you have probably already constructed line graphs as part of some science project. Here we will note some of the rules implicit in constructing the graph and will examine logarithmic graphs, which are often used to present nonlinear data. Graphs are prepared on lined graph paper, normally in pencil, though for permanent work ink is used.

1. Determine if the graph paper should be linear or depict a special kind of function (discussed later).
2. Arrange the data in tabular form for ease in plotting, and determine which variable is the dependent variable and which the independent variable. The dependent variable is normally plotted on the ordinate, with the independent on the abscissa. The chart title will refer to the dependent variable in terms of the independent variable.

3. Determine the approximate locations of the axes, keeping in mind that it is desirable to show the zero location for both variables. Normally the zero value is in the lower left-hand corner of the graph, unless positive and negative values are to be shown.
4. Determine the scale for each axis. Scale equals the range of the variable or data, divided by the scale length available. This number must be suitable for the graph paper available. Do not use awkward fractions or decimals. For instance, if the minimum grid size is 10 divisions to the inch, then the scale divisions should be 1, 2, 5, 10, or a multiple of these.
5. The axes should be marked with numerical values so the reader can easily determine the minimum scale. Also, you do not want the numbers on the axes to be overcrowded; again, this decreases the communicative function of the graph.
6. If the data come from theoretical or empirical equations, plot the information and draw a smooth curve through the points, so the points no longer appear. If the data are experimentally determined, the graph must show this and distinguish this from theoretical or empirical data. In this case the data are plotted as small dots with a circle around each one. The line joining the points should not pass through the circles, so the exact value of the data is clearly visible. If more than one set of data is displayed on the same chart, use different symbols to enclose the point—a triangle or square, for instance.
7. Both axes should be labeled with the appropriate variable description and the units of the variable, for instance, "Velocity, ft/sec," or "Temperature, °K." The axis label should be outside the axis and parallel to it. The title should include the names of the variables and succinctly and clearly denote the information on the graph. The reader should not have to read the text to determine the reason for the graph. Place the title so it does not interfere with reading the graph itself.

EXAMPLE 6.4

A thermometer has readings of degrees Celsius; plot the data with degrees Celsius as the ordinate and degrees Fahrenheit as the abscissa.

Degrees Celsius	Degrees Fahrenheit
0	32
16.7	62
33.3	92
50	122
66.7	152
83.3	182
100	212

SOLUTION

Given: The temperature readings of a thermometer in C and F.

Find: A plot of the data with degrees Celsius as the ordinate.

Assumptions: None.

Analysis: The graph paper available has five divisions per inch and the distance available for the ordinate is five inches and for the abscissa seven inches. For the abscissa the scale is $212/7 = 30.3°$/inch. In this case round the value to $30°$, which is divisible by five, yielding $6°$ per division. Notice that the entire range for the abscissa data was used to determine the scale, even though the lowest value is $32°$. For the ordinate the scale is $100/5 = 20°$/inch. In this instance the scale fits the division size easily with no adjustment necessary, yielding $4°$ per division. The graph is shown in Figure 6.7. A spreadsheet may be used to plot the data as shown in Figure 6.8.

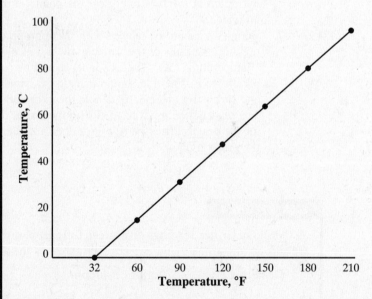

Figure 6.7 A line graph of the relationship of degrees Celsius to degrees Fahrenheit.

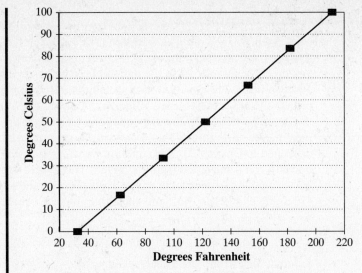

Figure 6.8 A spreadsheet of the relationship of degrees Celsius to degrees Fahrenheit; produced from data in Figure 6.7.

Often it is desirable to combine several graphs on the same chart, as Figure 6.9 illustrates. This is particularly useful in comparative analyses.

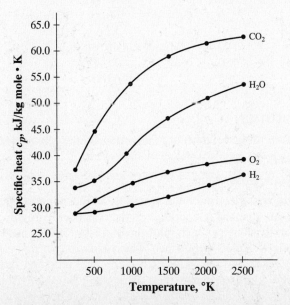

Figure 6.9 Multiple line graphs on the same chart, depicting the variation of specific heat with temperature for four gases or vapors.

EQUATION OF A STRAIGHT LINE

Very often you will want to determine the equation that a set of data yields, to show the empirical relationship between the variables. This is useful when you do not want to refer to the data table to get an ordinate value, given the abscissa value. The equation for a straight line is $y = mx + b$, where y is the ordinate value, x the abscissa value, m the slope, and b the y-intercept when x is zero. We can determine the equation of a straight line from the graphical information in Figure 6.5 as follows (a more sophisticated methodology is described in Chapter 7). The equation we use is $C = mF + b$, where m and b need to be determined from the graph. There were two unknowns, so pick any two points to determine m and b. Let us pick the following coordinates, (122, 50) and (212, 100), and substitute into the equation above.

$$50 = 122m + b$$
$$100 = 212m + b$$

Subtract one from the other, yielding $m = 0.5556$, and then substitute into either equation and solve for b, which yields $b = -17.78$. Finally, $C = 0.5556F - 17.78$. If the data were not linear, as we determined visually, then using a straight line to represent the data is inaccurate and misleading. For instance, the data for the boiling temperature versus pressure of water are nonlinear.

EXAMPLE 6.5

Determine the plot of the boiling temperature of water versus pressure for the following data:

Temperature (°C)	Pressure (kPa)
0	0.61
50	12.6
100	101.3
150	475.8
200	1553.8
250	3973

SOLUTION

Given: The variation of temperature with pressure for boiling water.

Find: Plot the data.

Assumptions: None.

Analysis: Determine the scale for the independent and dependent variables. In this case the temperature is the dependent variable. The graph paper has five divisions per inch, and there are five inches available for the ordinate: scale = 250/5 = 50°/inch. This yields 10° per division, as no adjustment is necessary in the scale size. For the pressure the scale is 3973/8 = 496.6 kPa/inch, where there are eight inches available for the abscissa. In this case, round up the scale value to 500 kPa/inch, yielding 100 kPa/division. This is a large value, considering some of the data. A plot of the data is shown in Figure 6.10. Figure 6.11 shows the spreadsheet plot of the data.

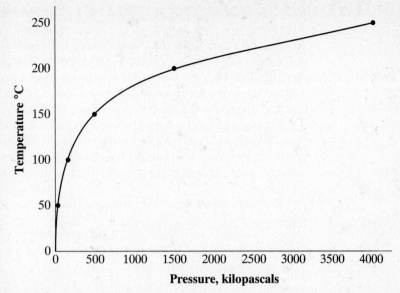

Figure 6.10 A line graph of the variation with pressure of boiling temperature water.

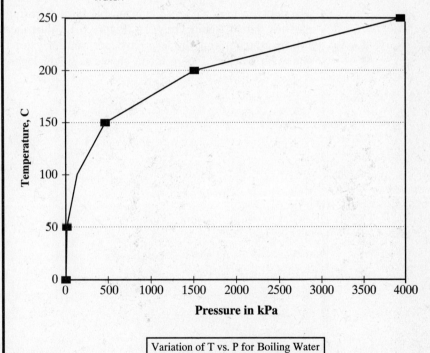

Figure 6.11 A spreadsheet of the variation with pressure of boiling temperature water; produced from data in Figure 6.10.

Comments: The data are nonlinear, and it is difficult using the graph to determine the temperature at low values of pressure. The next section shows a solution of this problem.

6.4 PLOTTING ON SEMILOG AND LOG-LOG PAPER

A variety of scales other than linear ones can be used for plotting data. Two of the most common involve the use of logarithms. Reasons for considering these are that the data may cover a large range and need to be compressed graphically, and a logarithmic plot may arrange nonlinear data in a straight line. Let's see why the log function causes the data to be compressed.

Semilog and log-log paper is based on logarithms to the base ten, rather than the natural logarithms that are more commonly used in engineering. A logarithm expresses any number by raising the number ten to a power. For instance, take the logs of the pressure data in the previous example.

Temperature (°C)	log (pressure kPa)
0	−0.21
50	1.09
100	2.00
150	2.68
200	3.19
250	3.60

The data are in much more manageable form. The numerical spread of the pressure data is significantly less. Figure 6.12 illustrates a plot of these data.

By using multiple-cycle graph paper, you can extend the range of the data you wish to plot. Figure 6.13a shows a single logarithm scale and Fig-

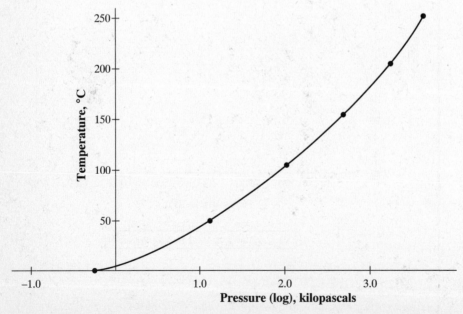

Figure 6.12 The plot of the log of pressure versus the boiling temperature for water.

6.4 PLOTTING ON SEMILOG AND LOG-LOG PAPER 149

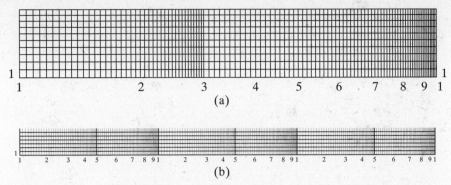

Figure 6.13 (a) A single-cycle logarithmic scale. (b) A three-cycle logarithmic scale.

ure 6.13b a three-cycle scale. The three-cycle scale covers a numerical range of one thousand. Figure 6.14 shows a three-cycle scale calibrated three ways, from 1 to 1000, from 10^3 to 10^6, and from 10^{-3} to 1. When using multiple-cycle graph paper, use the one with the fewest cycles possible to represent your data. The data is most clearly presented in this fashion.

How can you tell if your data should be represented by a semilog or log-log graph, and how does using logarithms produce a straight line from nonlinear data?

Consider the following equation:

$$y = be^{mx}, \qquad 6.1$$

and take the log of both sides, yielding

$$\log y = mx(\log e) + \log b. \qquad 6.2$$

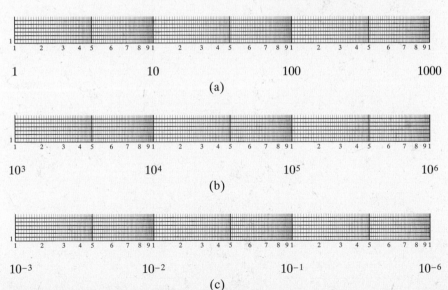

Figure 6.14 (a) A three-cycle scale calibrated from 1 to 1000. (b) A three-cycle scale calibrated from 1×10^3 to 1×10^6. (c) A three-cycle scale calibrated from 1×10^{-3} to 1.0.

In Equation 6.2 the log b and $m(\log e)$ are constant terms, so the equation is in the form of a straight line, with log y on one side and x on the other. Data that were described by Equation 6.1 would plot as a straight line on semilog

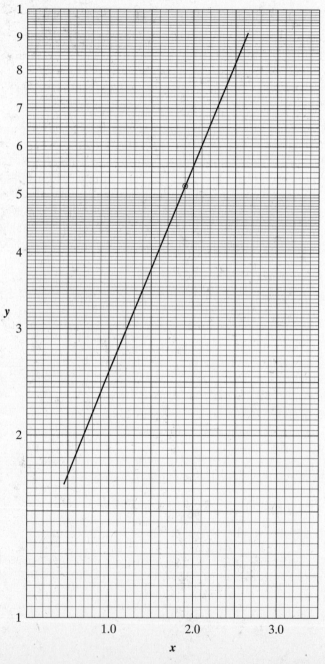

Figure 6.15 A graph of the equation $y = 1.3e^{0.7x}$ on semilog graph paper.

paper. One coordinate varies as a log and the other is linear. Figure 6.15 shows this function plotted on semilog paper.

In another common situation, the data are represented by a curve obeying the functional relationship

$$y = bx^m. \qquad 6.3$$

Take the logs of both sides, yielding

$$\log y = m(\log x) + \log b. \qquad 6.4$$

In Equation 6.4 m and $\log b$ are constants, but the two variables are represented as log values. This fits the form of a straight line, so Equation 6.3 is plotted as a straight line on log-log paper. Figure 6.16 illustrates a log-log plot of this function.

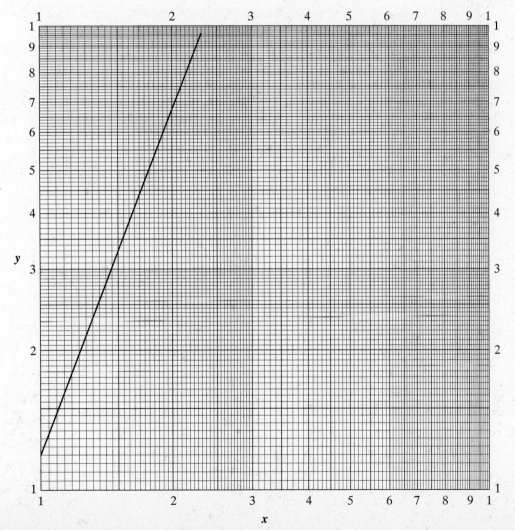

Figure 6.16 A log-log plot of the equation $y = 1.2x^{2.5}$.

EXAMPLE 6.6

Plot the following data on log-log paper and determine the equation relating x and y.

x	y
1	2.0
3	11.6
5	26.3
7	45.0
9	67.3

SOLUTION

Given: Data showing the variation of x and y.

Find: The plot of the data on log-log graph paper (Figure 6.17) and the equation of a straight line (Figure 6.18).

Sketch:

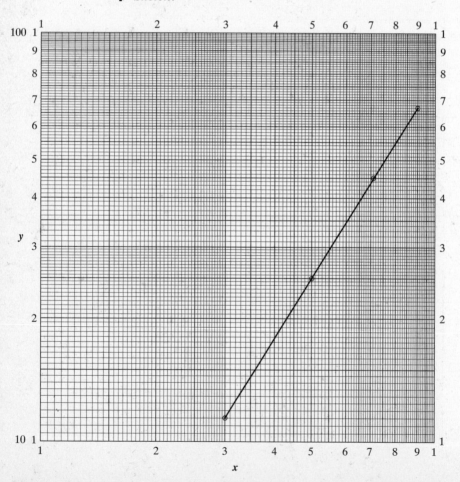

Figure 6.17 A log-log plot of the variation of x and y.

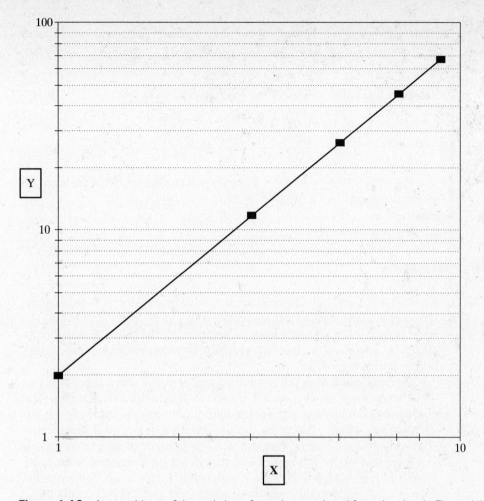

Figure 6.18 A spreadsheet of the variation of *x* and *y*; produced from the data in Figure 6.17.

Assumptions: None.

Analysis: To find the equation describing the data, substitute data from two points to solve for the equation of a straight line on the log-log plane. Equation 6.4 is used at the two points.

$$\log 26.3 = m(\log 5) + \log b$$
$$\log 67.3 = m(\log 9) + \log b$$

This yields upon evaluating the logs

$$1.420 = 0.699m + \log b$$

and

$$1.828 = 0.954m + \log b.$$

Subtract the first from the second, eliminating the log b.

$$0.408 = 0.255m$$
$$m = 1.6$$

Substitute the value of m in either equation and solve for b.

$$1.42 = 0.699 \times 1.6 + \log b$$
$$\log b = 0.3016$$
$$b = 2.00$$

The equation is $y = 2.0x^{1.6}$. To verify that this is true, check the equation at another point, $y = 2.0(3.0)^{1.6} = 11.6$.

6.5 SKETCHING

A sketch is not a sloppy drawing, nor is it inaccurate; rather it is an accurate, freehand drawing used to transmit information visually. To sketch well you need, in addition to the techniques presented here, the ability to visualize an object. A course in engineering graphics or engineering drawing is very helpful in this regard.

The only equipment you need are a pen or pencil and a sheet of paper. When you want to sketch in pencil, a medium-weight pencil is best. You want one that will not smudge but is not so hard it prevents you from varying line intensity and width. Most often you will sketch a device so it looks three-dimensional. An isometric pictorial drawing is particularly useful, and we will cover that briefly. It is convenient, though not a requirement, to use paper printed with an isometric grid pattern, shown in Figure 6.19. This paper has lines at a 30° angle to the horizontal, which help give the picture depth.

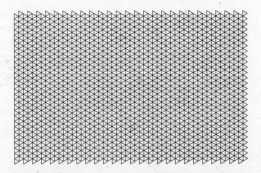

Figure 6.19 An isometric grid pattern.

Let's consider the techniques used to draw an elementary sketch.

STEP 1. Lay out the three-dimensional grid, keeping lines parallel and in this case vertical lines vertical, but with what would be the x and y coordinates slanted at 30° (Figure 6.20). It is possible to use 45° as well, though the slope is a little dramatic. The construction lines should be light and thin, as you will erase them when the drawing is completed. The model should fit entirely within this frame.

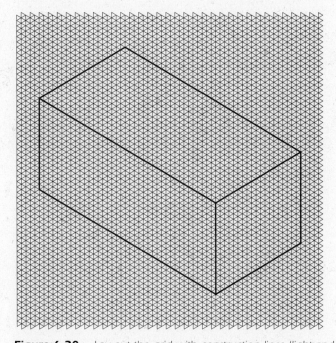

Figure 6.20 Lay out the grid with construction lines (light and thin).

STEP 2. Draw the object you have in mind, remembering the parallel rule for lines. Also, circles are not circles, but ellipses in isometric sketches (Figure 6.21).

STEP 3. Remove the construction lines. Some feel this is not a necessary step, but the visual difference between Figures 6.21 and 6.22 seems important.

156   6/GRAPHICAL COMMUNICATION

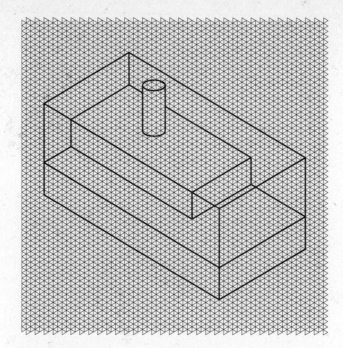

Figure 6.21   Draw the lines of the object.

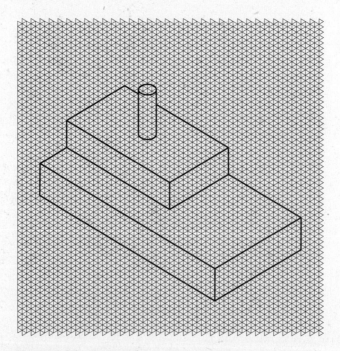

Figure 6.22   Remove the construction lines.

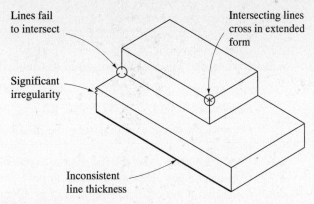

Figure 6.23 Sketching errors.

Some additional errors that detract from the appearance of your sketch are shown in Figure 6.23. Of course the errors are exaggerated, but they certainly have a negative impact. Have all lines intersect and terminate where they should, not forming crosses or other shapes that are distracting and, in the worst situation, misleading. Equally bad is to have the lines not meet at all, or some of the lines not meet with a common terminus. Sketches are freehand drawings, so some irregularity in line direction is expected, but significant irregularities should be corrected. Visible lines should be dark and of uniform thickness: this may mean sharpening your pencil occasionally. The thick line shown in Figure 6.23 detracts from the sketch.

REFERENCES

1. Earle, J. H. *Engineering Design Graphics.* 4th ed. Addison-Wesley, Reading, MA, 1983.
2. Spence, W. P. *Engineering Graphics.* Prentice-Hall, Englewood Cliffs, NJ, 1985.
3. Voland, G. G. S. *Modern Engineering Graphics and Design.* West, St. Paul, 1987.

PROBLEMS

6.1. Construct a circle chart depicting the following grade distribution data for a class: 7 A's, 12 B's, 21 C's, 10 D's, and 4 F's.

6.2. Construct a circle chart indicating the distribution of engineering majors at a university given the following information: chemical engineering—210 students; civil—355; electrical—689; industrial—310; and mechanical—474.

6.3. Construct a circle chart illustrating a state's annual expenditure on public roads. Approximate annual figures are: construction—$14 million; maintenance—$8 million; equipment purchase and maintenance—$3.5 million; bonds—$6.2 million; engineering and administration—$2.1 million.

6.4. An alloy has the following composition: lead—24%; tin—16%; aluminum—16.6%; copper—4.6%; zinc—38.8%. Create a circle chart for the alloy.

6.5. A company has the following age distribution among its employees:

Age group	<20	20–29	30–39	40–49	50–59	>59
Employees	15	71	82	68	51	19

Construct a circle chart showing the distribution by age of the employees.

6.6. Construct a bar chart for the data in problem 6.1.
6.7. Construct a bar chart for the data in problem 6.2.
6.8. Construct a bar chart for the data in problem 6.3.
6.9. Construct a bar chart for the data in problem 6.4.
6.10. Construct a bar chart for the data in problem 6.5.
6.11. A manufacturing company has the following annual expenditures: research and development—$21 million; engineering design—$11 million; tooling and equipment—$15 million; labor—$33 million; and marketing/distribution—$20 million. Construct a bar chart illustrating these expenditures.

6.12. A manufacturing company produces two models of a milling machine and has the following weekly production forecast.

Year	Model A	Model B
1989	550	700
1990	600	825
1991	650	850
1992	500	950
1993	400	1100
1994	0	1800

Construct one bar chart showing the weekly production forecast for both models.

6.13. The following data were obtained from a laboratory experiment where a structure was subjected to a force, causing it to deflect.

Force (kN)	Deflection (mm)
0	0
100	0.3
300	1.8
500	2.9
700	4.1
900	5.2
1100	6.3

Construct a line graph of the data and determine the equation of a straight line passing through the points.

6.14. The following data represent the weight per foot of wire rope for various wire diameters. Construct a graph of the data.

Diameter (in)	Weight (lbf/ft)
$\frac{3}{4}$	1.41
1	2.50
$1\frac{1}{4}$	3.91
$1\frac{1}{2}$	5.63
$1\frac{3}{4}$	7.66
2	10.00
$2\frac{1}{4}$	12.50
$2\frac{1}{2}$	15.2
$2\frac{3}{4}$	18.3
3	22.2
$3\frac{1}{2}$	29.9
4	38.4

6.15. The following data represent the standard temperature and pressure variation with altitude. Plot both temperature and pressure variation on the chart.

Altitude (ft)	Temperature (°F)	Pressure (psia)
0	59.0	14.69
5,000	41.2	12.22
10,000	23.3	10.10
15,000	5.5	8.29
20,000	−24.6	6.75
25,000	−30.2	5.45
30,000	−48.0	4.36
35,000	−65.8	3.46
36,089	−69.7	3.28
40,000	−69.7	2.72
50,000	−69.7	1.68

6.16. A diesel engine has the following horsepower outputs as a function of a percentage of its rated rpm. Two outputs are given for 80% throttle and 100% throttle. Plot both curves on the same chart.

Percentage-rated rpm	80% (hp)	100% (hp)
40	30	40
60	52	66
80	70	86
100	80	100

6.17. The bending moment of a shaft supported by bearings is

$$M = wL^2/12,$$

where M is the bending moment in inch-lbf, L is the span between bearings in inches, and w is the weight of a unit length of shaft, lbf/inch. Plot the bending moment for a shaft with $w = 15$ lbf/inch for the span varying from 6 inches to 2 feet in 2-inch intervals.

6.18. A capacitor is discharged and the following table gives the values of the charge in coulombs in the capacitor with time. Plot this data.

Time	Charge (C)
0	25×10^{-6}
0.1	16.8×10^{-6}
0.2	11.3×10^{-6}
0.3	7.5×10^{-6}
0.4	5.0×10^{-6}
0.5	3.5×10^{-6}
0.6	2.3×10^{-6}
0.7	1.5×10^{-6}
0.8	1.0×10^{-6}
0.9	0.8×10^{-6}

6.19. The following table indicates the temperature-time history of a metal ingot when it was immersed in boiling water. Plot the information and determine the equation of a straight line describing the data points.

Time (sec)	Temperature (°F)
0	−10
5	5
10	20
15	35
20	50
25	65
30	80

6.20. A group of machine parts, made of several components, were tested repetitively under a variety of loads until they failed. The following data indicate the loads and the cycles required until failure occurred. Plot the data on semilog paper and determine the equation relating cycles to failure as a function of load.

Load (lbf)	N (cycles of failure)
10	4.10×10^{6}
15	2.74×10^{6}
20	1.96×10^{6}
25	1.40×10^{6}
30	1.00×10^{6}
35	7.18×10^{5}
40	5.14×10^{5}

6.21. For the data presented in problem 6.14, determine the log-log plot and the equation describing the weight in terms of the diameter.

6.22. Plot the pressure-altitude data from problem 6.15 on semilog paper and determine the equation for pressure as a function of altitude.

6.23. Plot the data from problem 6.17 on log-log paper.

6.24. Plot the data from problem 6.18 on semilog paper and determine the equation for charge remaining as a function of time.

6.25. Plot the function $y = 1.9x^{2.1}$ on rectangular graph paper and on log-log paper.

6.26. Plot the function $y = 1.9x^{-2.1}$ on rectangular graph paper and on log-log paper.

6.27. Plot the function $y = 2.3e^{1.5x}$ on rectangular graph paper and on semilog paper.

6.28. Plot the function $y = 2.3e^{-1.5x}$ on rectangular graph paper and on semilog paper.

6.29. A gear set in a laboratory can lift a certain amount of weight, given in the following table. Determine the equation that best describes the data.

Applied force (lbf)	Weight lifted (lbf)
15	184
29.3	286.4
35.4	334.1
46.4	375
55.9	463.6

6.30. Determine the equation that best describes the data.

X	Y
1	3
2	2.27
3	1.93
4	1.72
5	1.57
6	1.46

6.31. Determine the empirical equation that best fits the data.

X	Y
1	80.3
1.5	270
2.0	1,614
2.5	7,232
3.0	32,412

6.32. Draw an isometric sketch of a two-inch cube five times. Each version should have one of the defects, only once, as shown in Figure 6.23, and one should be correct. Look at the sketches and see how the defect detracts from the drawing, hence from your communication using the sketch.

6.33. Draw sketches (a) through (f) yourself. Notice that it is much easier to sketch when the grid lines of the graph are included.

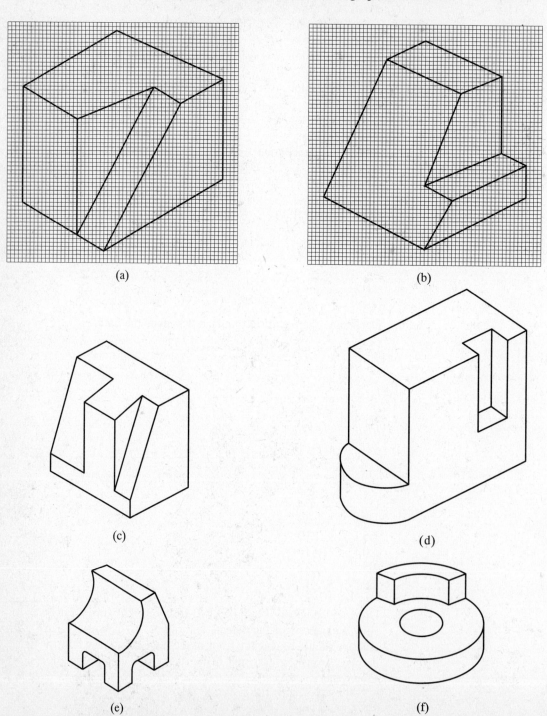

CHAPTER 7

Statistics and Error Analysis

CHAPTER OBJECTIVES

To learn the measures of central tendency of data.

To learn the measures of data dispersion.

To become familiar with probability theory and normal distributions.

To apply linear regression analysis.

To understand error analysis and error propagation.

(USDA/SCS Photo.)

7/STATISTICS AND ERROR ANALYSIS

Engineering students are often involved with statistics as part of their educational program, sometimes in course work, frequently in labs. Engineers in industry use statistics and the results of statistical analysis as part of their work function.

7.1 INTRODUCTION

Statistics is the collection, organization, and interpretation of numerical data. The data may be from experiments run in a laboratory or from economic information. Daily we are deluged with a variety of statistical information, ranging from the weather forecasts, to inflation rates, to the predicted outcome of political elections. It is important that we be able to interpret this information wisely as statistics can also easily befuddle us at times. Moreover, statistics gives the engineer a means of quantifying uncertainty, a constant companion in engineering.

In industry you will use statistical analysis in a variety of circumstances. For instance, in a manufacturing plant the quality control engineer must sample and test a certain number of the components to assure that the manufacturing process and product are meeting the expected quality standards. Then the engineer must judge from the measurements on a few whether the

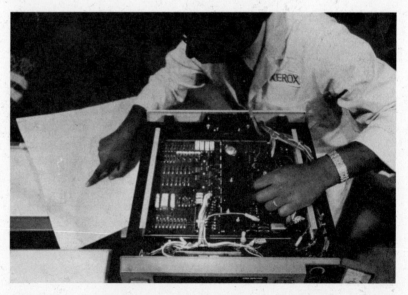

This engineer is checking performance to make sure it meets the design operating specifications. In a quality assurance program a certain number of units are randomly sampled and checked for their performance. This subgroup has been mathematically shown to be representative of the whole and is used as a predictor of the group quality. (Photo Courtesy of Xerox Corporation.)

majority are within bounds. Obviously, this is a complex area of study; people spend their entire careers specializing in statistics, and we cannot hope to delve into all the complexities of statistics here. However, an understanding of fundamentals will help you perform better as engineering students and as practicing engineers.

7.2 MEASURES OF CENTRAL TENDENCY

When we have collected a set of numerical data, we most often want to characterize it so we can visualize the cumulative nature of data, rather than the entire set. Most frequently we look at the average value of the data, the central tendency of the data. The average that we most often have in mind is the arithmetic mean, $\bar{x}$, which is defined as

$$\bar{x} = \sum_{i=1}^{n} \frac{x_i}{n} \qquad 7.1$$

where $\bar{x}$ is the average value, x_i is the individual values that are to be averaged, and n is the total number of x_i's. The symbol Σ represents summation. Let's see how this is used. Find the arithmetic mean of the numbers 10, 8, 12, 5, 7, 9, and 9.

$$\bar{x} = \sum_{i=1}^{7} \frac{x_i}{n} = \frac{10 + 8 + 12 + 5 + 7 + 9 + 9}{7} = \frac{60}{7} = 8.6$$

The mean gives us an understanding of the centrality of the available data group, but the results can be misleading if there are some data that are not characteristic of the group. In that case the median is better used to determine the centrality. The median is found by arranging the data in order of magnitude from smallest to largest. The value at the halfway point, if the number of data is odd, is the median value. If the number of data is even, then the average of the middle two values is the median value of the data.

EXAMPLE 7.1

The following data represent the prices of new automobiles purchased by your neighbors in the past two years. Determine the mean and median values: $15,000, $17,000, $11,000, $13,000, $43,000, $15,000, $12,000.

SOLUTION

Given: Various prices for automobiles.

Find: The mean and median values.

Assumptions: None.

Analysis: The mean value is

$$\frac{126,000}{7} = \$18,000.$$

Arranging the data in ascending value yields $11,000, $12,000, $13,000, $15,000, $15,000, $17,000, $43,000. The median value is $15,000. In this case, because one car cost $43,000, an anomaly compared to the rest of the data, the median is more indicative of the centrality of the data.

There are other measures of the central tendency of the data. The mode can be useful at times, particularly when looking at data distributions. A mode occurs when a piece of data repeats itself; for instance, in the car price data, the value of $15,000 occurred twice, so the data have one mode. If no value is repeated, the data have no modality, and if different values repeat, then there are two, three, or more modes associated with the data.

The following numbers have the same mean and median, but different modes:

2, 6, 4, 3, 9 no modes

7, 7, 4, 3, 3 two modes (bimodal).

We do not have to adopt just one measure but can use our common sense and multiple measures to characterize the data.

A final measure that we will analyze is the weighted arithmetic mean, $\bar{X}$, which is used when certain data values have more importance (weight) than others. We use weighting factors in Chapter 7 in discussing multicriteria decision making. In this case the weighting factors are $w_1, w_2, w_n, \ldots$, and associated with numbers $X_1, X_2, X_n, \ldots$. The weighted arithmetic mean is

$$\bar{X} = \frac{w_1 X_1 + w_2 X_2 + w_3 X_3 + \cdots + w_n X_n}{w_1 + w_2 + w_3 + \cdots + w_n} = \sum_{i=1}^{n} \frac{w_i X_i}{\Sigma w_i}. \qquad 7.2$$

EXAMPLE 7.2

Given the following data with their weighting factors, determine weighted arithmetic mean.

Data (x)	Weighting factor (w)
10	1
12	1.5
20	1
15	2
19	3
16	1

SOLUTION

Given: A set of data and its weighting factors.

Find: The weighted arithmetic mean for the data.

Assumptions: None.

Analysis: The weighted arithmetic mean is

$$\bar{X} = \frac{1 \times 10 + 1.5 \times 12 + 1 \times 20 + 2 \times 15 + 3 \times 19 + 1 \times 16}{1 + 1.5 + 1 + 2 + 3 + 1}$$

$$= 15.9.$$

7.3 MEASURES OF DISPERSION

Just as a set of numbers has a centrality, so it has a dispersion. The scattering of the data about a central point provides us with additional information about the set. Consider the number set (10, 10, 10, 10, 10), which has a mean of 10, and another set with the same mean, (1, 20, 14, 5, 10). The mean does not indicate the scattering of the second group of data.

Several indicators can be used to indicate the scattering of the data. The range, R, is defined as the difference between the largest and smallest values of the data. Thus, for the two data sets above the range of the first set is 0 and of the second set 19.

An individual deviation is the difference between any data point and the mean. Note that the sum of the individual deviations is zero: the individual deviations may be positive or negative, depending on whether or not any given data point is greater than or less than the mean, and since they are equally dispersed about the mean their sum is zero.

A composite deviation would be useful, and there are ways to overcome the problem of the deviations' sum being zero. We could use the absolute value of the deviations and sum these: this is the mean deviation. A preferred method is the standard deviation. The standard deviation is also used with the normal distribution curve in probability theory (Section 7.4) and hence becomes the most common and useful dispersion correlator.

The standard deviation, σ, is defined as

$$\sigma = \sqrt{\sum_{i=1}^{n} \frac{(x_i - \bar{x})^2}{n}}. \qquad 7.3$$

EXAMPLE 7.3

Determine the range and standard deviation for the following set of numbers: 5, 8, 10, 15, 12, 18, 11, 7.

SOLUTION

Given: A set of numbers.
Find: The range and standard deviation of the data set.
Assumptions: None.

Analysis: The range is $R = 18 - 5 = 13$. The mean is

$$\bar{x} = \sum_{i=1}^{n} x_i/n = 86/8 = 10.75.$$

x_i	$(x_i - \bar{x})^2$
5	33.06
8	7.56
10	0.56
15	18.06
12	1.56
18	52.56
11	0.06
7	14.06

$$\sum_{i=1}^{n}(x_i - \bar{x})^2 = 127.48$$

$$\sigma = \sqrt{127.48/8} = 3.99$$

7.4 PROBABILITY AND NORMAL DISTRIBUTION

Very often when we think of statistics and probability we think chances: what is the chance that something will happen? Probability helps in estimating chances. Consider flipping a coin: what is the chance, the probability, that it will be heads when it lands? The probability of an event occurring is defined as the frequency of the occurrence divided by the number of attempts, letting the number of attempts be quite large. In engineering probability is important in sampling of parts for quality control; the quality of the sampled parts should relate to that of the remaining parts, and should be the same if the sample is properly selected. Thus probability theory and statistics are closely allied.

Returning to the coin, the probability is

$$p = s/n = 1/2, \qquad \qquad 7.4$$

where s is the number of desired outcome (in this case there is one outcome, heads), and n is the number of possible outcomes of an event (in this case two, heads or tails). The probability of all other events, q, is

$$q = (n - s)/n. \qquad \qquad 7.5$$

If we combine Equations 7.4 and 7.5 we find that the sum is one, or 100 percent probability that some event will occur. Thus, when the probability is one, the event is certain to occur, and when the probability is zero the event cannot occur.

Any time you flip a coin the probability is 50 percent that it will come up heads, regardless of how many times you have flipped it before. Say you are matching coins with a friend. You both want to have heads occur simultaneously. The probability for you getting heads is 1/2, the probability for your friend is 1/2, and the combined probability is the product of the two, or 1/4, 25 percent. To generalize this as a law, if the probability of m independent events is $p_1, p_2, p_3, \ldots, p_m$, then the probability that they will all occur simultaneously, P, is

$$P = p_1 p_2 p_3 \ldots p_m. \qquad 7.6$$

Now let's consider another situation. You have neatly placed ten pennies on a table, near the edge, heads up. Your younger brother comes along and sweeps them off the table. What is the probability that they will all be heads up when they land on the floor? $P = (1/2)^{10} = 1/1024$, or about one chance in a thousand. Let's change the situation to that of at least one being heads up. In this case the law in Equation 7.6 no longer applies, as we are no longer asking that all looked-for events happen simultaneously. The probability, P, that one of the looked-for events will occur (one head) is

$$P = 1 - \text{(probability of all tails)}$$

$$P = 1 - \frac{1}{1024} = 0.999. \qquad 7.7$$

In both of these cases the events must be mutually independent: one event does not affect the other events. If it did, then the probability function cannot be as readily determined, and Equations 7.6 and 7.7 are not valid.

The frequency distribution of an event is often important in laboratory situations when you measure a quantity that is supposedly constant, but actually varies somewhat.

EXAMPLE 7.4

An engineer is gathering data on a power plant. At a certain place in the power plant the steam pressure is supposedly constant, but actually varies slightly. There were 100 observations, yielding the following results. Plot a histogram of the data.

Pressure (psia)	Number of results (n)
397	1
398	3
399	11
400	26
401	33
402	17
403	6
404	3

SOLUTION

Given: A data set of steam pressure observations.

Find: The histogram plot of the data.

Assumptions: None.

Sketch:

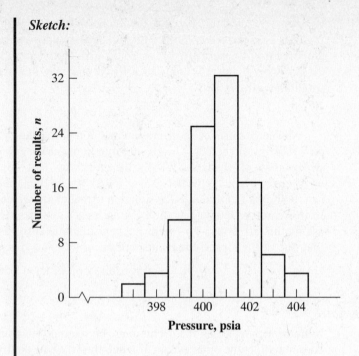

Figure 7.1 A histogram for the data in Example 7.4.

Comments: The histogram gives a frequency distribution of the data.

The shape of the histogram can be greatly altered by changing the number of intervals. For instance, if we chose three intervals, the data would have a distribution as shown in Figure 7.2.

Pressure	Number of results
398–400	15
401–402	76
403–404	9

Clearly, the implications are different. Guidelines help in selecting the best interval size for plotting the data. One of these is called the Sturgis rule:

$$I = 1 + 3.3 \, (\log n), \qquad 7.8$$

where I represents the number of intervals and n is the number of observations. For this situation $I = 1 + 3.3 \, (\log 100) = 7.6$, or rounded off to 8, the number chosen for the example.

The results of Example 7.4 still may not be quite what you have in mind regarding frequency distribution. When the number of observations is infinite, the results will form a smooth curve. One of the most common distribution curves is the bell-shaped curve, or normal distribution curve,

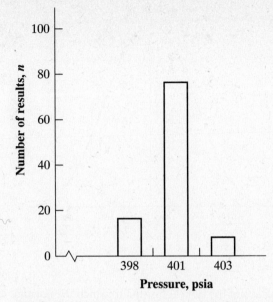

Figure 7.2 A histogram for consolidated data in Example 7.4.

shown in Figure 7.3 (Discussion of other distribution curves is beyond our scope.) The equation describing the curve in Figure 7.3 is

$$Y = \frac{1}{\sigma\sqrt{2\pi}}[e^{-(x-m)^2/2\sigma^2}], \qquad 7.9$$

where Y is the number of results (the frequency), x is the magnitude of the measured event (the pressure), m is the mean value of the events, and σ is the standard deviation.

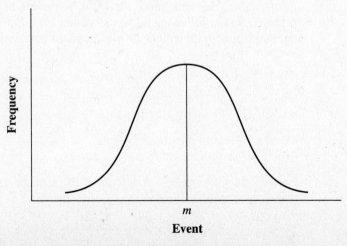

Figure 7.3 A normal distribution curve.

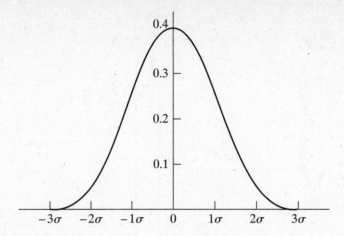

Figure 7.4 A standardized normal distribution curve.

This is not the typical fashion in which the normal, or Gaussian, distribution curve is given, as the variables may be changed to make the standard normal distribution curve. To do this let

$$y = Y\sigma \quad \text{and} \quad z = (x - m)/\sigma. \qquad 7.10$$

Substitution into Equation 7.9 yields

$$y = \frac{1}{\sqrt{2\pi}}(e^{-z^2/2}). \qquad 7.11$$

A plot of Equation 7.11 is shown in Figure 7.4. A tremendous advantage to using this curve is that the abscissa is in multiples of the standard deviation. Furthermore, the area under the curve integrates to unity, or 100 percent probability. The following table, calculated from Equation 7.10, shows why the abscissa is in increments of the standard deviation.

x	z
$m + 2\sigma$	$+2$
$m + 1\sigma$	$+1$
m	0
$m - 1\sigma$	-1
$m - 2\sigma$	-2

Figure 7.5a shows that 68.3 percent of the area, or data, lies within plus or minus one standard deviation of the mean ($z = 0$); Figure 7.5b that 95.5 percent of the area lies within plus or minus two standard deviations of the mean; and Figure 7.5c that 99.7 percent of the area lies within plus or minus three standard deviations of the mean.

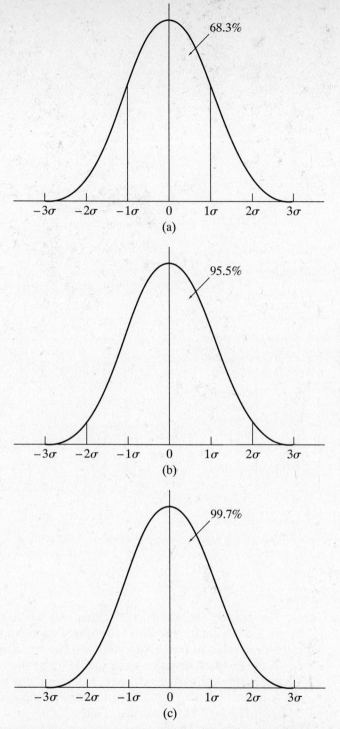

Figure 7.5 (a) 68.3% of the data are within ±σ of the mean based on a standardized normal distribution curve. (b) 95.5% of the data are within ±2σ of the mean based on a standardized normal distribution curve. (c) 99.7% of the data are within ±3σ of the mean based on a standardized normal distribution curve.

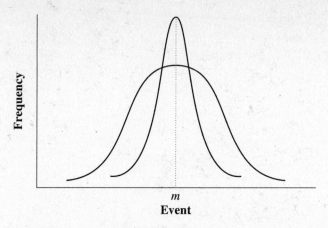

Figure 7.6 Two normal distribution curves with the same mean value, but different standard deviations (σ_2, σ_1).

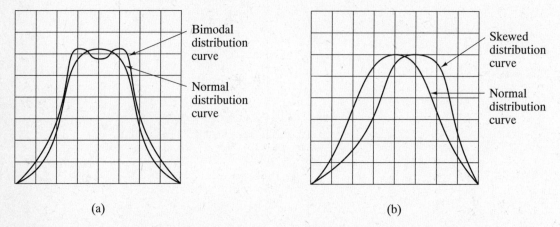

Figure 7.7 (a) A bimodal distribution curve and a normal distribution curve. (b) A skewed distribution curve and a normal distribution curve.

The smaller the standard deviation, the closer the data fit together. Figure 7.6 illustrates this with two normal distribution curves, having the same mean value, m, but looking entirely different. In the narrow distribution curve, the standard deviation has a smaller value than in the broader curve. Thus the probability of being distant from the mean is less.

A note of caution. Not all distributions are normal ones; some may be bimodal, Figure 7.7a, or skewed, Figure 7.7b. These frequently occur in grade distributions, among other situations, and applying a normal distribution function to these may lead to erroneous conclusions. Don't assume the data set you are analyzing follows the Gaussian distribution—plot it to be sure.

7.5 LINEAR REGRESSION ANALYSIS

In Chapter 6 it was mentioned that a more sophisticated means, linear regression analysis, could be used to determine the equation of a straight line. Perhaps the data from an experiment or a series of observations produce, when plotted, a scatter diagram as shown in Figure 7.8. The data are from an experiment on measuring voltage and current.

Voltage (x)	Current (y, milliamperes)
2	12
3	20
4	24
6	35
7	38
8	45
10	53

What we would like to determine is the equation of a straight line that best represents the data.

You could take a transparent straight edge and align it as closely as possible with the data, draw a line, and determine the equation of this line by

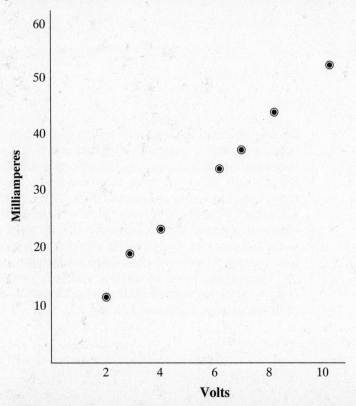

Figure 7.8 A scatter diagram of the relationship between volts and milliamperes.

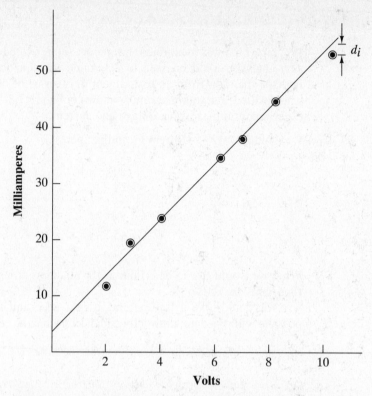

Figure 7.9 A straight line, $y = a + bx$, drawn through the data points from Figure 7.8. d_i = deviation of the data from the line at point $x = 10$.

the slope-intercept method. You want the deviations from the line to any point to be a minimum, considering all the data points. It is possible to perform this task mathematically. Figure 7.9 graphs the data with a straight line and shows the deviation, d_i, at one point. The deviation is the difference in y values between the line's value and the data point at a given value of x. What we want to do mathematically is to make the sum of the squares of the deviations a minimum. Squaring the deviations prevents the positive and negative deviations from canceling each other and prevents a line, not fitting the data at all, from meeting the constraint of minimum sum deviation.

The mathematics for determining the coefficients of the line, $y = a + bx$, is not complex but requires differential calculus. The results for any data set (x, y) are given in Equations 7.12 and 7.13.

$$b = \frac{n \sum_i x_i y_i - \sum_i x_i \sum_i y_i}{n \sum_i x_i^2 - \left(\sum_i x_i\right)^2} \quad \quad \textbf{7.12}$$

$$a = \frac{\sum_i y_i - b \sum_i x_i}{n} \quad \quad \textbf{7.13}$$

The data from Figure 7.8 result in the following values for a and b.

x	y	x^2	xy
2	12	4	24
3	20	9	60
4	24	16	96
6	35	36	210
7	38	49	266
8	45	64	360
10	53	100	530
$\sum_i x_i = 40$	$\sum_i y_i = 227$	$\sum_i x_i^2 = 278$	$\sum_i x_i y_i = 1546$

$$b = \frac{(7)(1546) - (40)(227)}{(7)(278) - (40)^2} = \frac{1742}{346} = 5.035$$

$$a = \frac{(227) - (5.035)(40)}{7} = \frac{25.6}{7} = 3.657$$

The equation of the straight line is $y = 3.657 + 5.035x$.

7.6 ERROR ANALYSIS AND ERROR PROPAGATION

We have been assuming that the measurements, with resulting data, are correct in our problems and examples. Any readings that are made involve error, however. Systematic errors are associated with the instrument or the technique used in the measurement. For instance, a thermometer reads 102°C rather than 100°C for the boiling point of water. A yardstick is not exactly 36 inches long, and the inch markings are not exactly one inch. In looking at a gage you must interpret where the needle is pointing; a systematic error is associated with reading the value from the instrument. Random errors are produced by a wide variety of unpredictable variations in the experiment. Perhaps fluctuations in atmospheric pressure or opening and closing the laboratory door causes variations in the experimental results. These random errors cannot be duplicated. A third kind of error is not an error at all but a blunder. Perhaps you made a mistake in reading an instrument or didn't calibrate a piece of test apparatus, and the results are in error. Ideally, the experiment should be rerun.

The terms *accuracy* and *precision* refer to the two types of errors that will occur in any experiment. High accuracy refers to a measurement that has a small systematic error. A micrometer, measuring to one-thousandth of an inch, is used to measure thickness, rather than a ruler that measures to the nearest thirty-second of an inch. An experiment with high precision is one in which the random errors are small, and the experiment is repeatable. The experimental design should minimize errors, by controlling extraneous events to minimize random errors and by using accurate instrumentation and measurement techniques to minimize systematic errors. In general you will not have control over the instrumentation that is used in experiments in engineering and physics, but you can be aware of the errors that are created by use of the existing instrumentation.

The most important aspect of error analysis is calculating the effect that individual errors in experimental observation have on the final result. Thus, by knowing the error in a temperature reading we can determine the error in a calculation using the temperature. The percent error in any reading is the uncertainty of the reading divided by its nominal value. A pipe's diameter is measured and found to be 4 centimeters, within 0.008 centimeters, so the percent error associated with the measurement is $(0.008/4) \times 100 = 0.2\%$. Consider that you are measuring a cylinder's volume by determining its diameter and height and then calculating the volume. Let Δr and Δh represent the error in the measurement of the radius, r, and height, h. The initial volume is

$$V = \pi r^2 h,$$

and the volume with the measurement errors is

$$V + \Delta V = \pi (r + \Delta r)^2 (h + \Delta h). \qquad 7.14$$

Thus,

$$V + \Delta V - V = \pi (r^2 \Delta h + 2rh\Delta r + \Delta r^2 h + 2r\Delta r \Delta h + \Delta r^2 \Delta h); \qquad 7.15$$

neglecting higher-order terms as being too small compared to the remaining terms, Equation 7.15 becomes

$$\Delta V = \pi (r^2 \Delta h + 2rh\Delta r). \qquad 7.16$$

If we divide by V to obtain the error in the total volume, the result is

The control room in a nuclear power plant provides the operators with information concerning all aspects of the plant operating condition. (Courtesy of Trident Engineering Associates)

7.6 ERROR ANALYSIS AND ERROR PROPAGATION

$$\Delta V/V = \frac{\pi(r^2\Delta h + 2rh\Delta r)}{\pi r^2 h} = \frac{\Delta h}{h} + \frac{2\Delta r}{r}. \qquad 7.17$$

For instance, if the radius is 50 cm and $\Delta r = 0.5$ cm, and the height is 100 cm and $\Delta h = 0.5$ cm, then the percent error in the volume is

$$\frac{\Delta V}{V} = \frac{0.5}{100} + \frac{(2)(0.5)}{50} = 0.025 = 2.5\%.$$

This can be extended to more complicated equations. I will not derive the general error equation, but will demonstrate its use in the following example. Let us determine the percent error in Y, given by the equation below,

$$Y = AB^m/C^n, \qquad 7.18$$

where A, B, C are measured quantities and m and n are the exponents of these variables. The percent error in Y is

$$\Delta Y/Y = \sqrt{\left(\frac{\Delta A}{A}\right)^2 + \left(\frac{m\Delta B}{B}\right)^2 + \left(\frac{n\Delta C}{C}\right)^2}. \qquad 7.19$$

Notice that the effect of the negative sign on n is not important, as the value of $\Delta C/C$ is squared.

EXAMPLE 7.5

The period of a pendulum is given by the equation $t = 2\pi(L/g)^{1/2}$ In a physics lab the following measurements were made: $t = 2$ seconds within 0.02 seconds; the length of the pendulum is 1 meter within 0.01 meters. Determine the percent error in the gravitational acceleration, and the values of the gravitational acceleration and the error.

SOLUTION

Given: The equation for the period of a pendulum, the pendulum's length and period and the error associated with each measurement.

Find: The gravitational acceleration, its percent error and numerical error.

Sketch and Given Data:

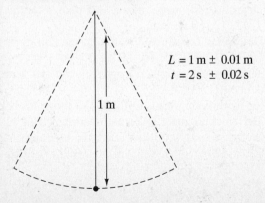

Figure 7.10

Assumptions: None.

Analysis: Solve the equation for a pendulum's period for the gravitational acceleration, yielding

$$g = 4\pi^2 L/t^2. \qquad (a)$$

This is in the form of Equation 7.18. The percent error may be determined from Equation 7.19 as

$$\Delta g/g = \sqrt{\left(\frac{2\Delta t}{t}\right)^2 + \left(\frac{\Delta L}{L}\right)^2} = \sqrt{\left(\frac{(2)(0.02)}{2}\right)^2 + \left(\frac{0.01}{1}\right)^2}$$

$$\Delta g/g = 0.022 = 2.2\%.$$

Solve Equation (a) for g, yielding

$$g = \frac{4\pi(1 \text{ m})}{(2 \text{ s})^2} = 9.869 \text{ m/s}^2.$$

Substitute into the equation for percent error to determine the numerical error in g.

$$\Delta g = (0.022)(9.869 \text{ m/s}^2) = 0.217 \text{ m/s}^2$$

REFERENCES

1. Beckwith, T. G.; Buck, N. L.; and Marangoni, R. D. *Mechanical Measurements.* 3d ed. Addison-Wesley, Reading, MA, 1982.
2. Neville, A. M., and Kennedy, J. B. *Basic Statistical Methods for Engineers and Scientists.* International Textbook, Scranton, PA, 1964.
3. Walpole, R. E., and Myers, R. H. *Probability and Statistics for Engineers and Scientists.* 4th ed. Macmillan, New York, 1989.
4. Young, H. D. *Statistical Treatment of Experimental Data.* McGraw-Hill, New York, 1962.

PROBLEMS

7.1. Determine the range, mean, and median for the following numbers: 9, 7, 12, 13, 10, 9, 8.

7.2. Determine the range, mean, and median for the following numbers: 15, 16, 12, 17, 15, 10, 11.

7.3. The following numbers represent the test scores of a group of students: 82, 75, 64, 86, 77, 81, 68, 94, 93, 65, 49, 77, 64, 55. Determine the arithmetic mean, median, and mode for the grades.

7.4. The chief engineer at a small manufacturing plant is determining the average monthly utility bill for the operations. The monthly bills are $200, $245, $175, $325, $210, $390, $295, $215, and $255. Determine the mean and median.

7.5. The number of cars crossing an intersection is measured for 15-minute intervals at various times of day, yielding the following data.

Time of day	Cars per 15 minutes
12 midnight to 6:00 A.M.	10
6:00 A.M. to 10:00 A.M.	78
10:00 A.M. to 1:00 P.M.	45
1:00 P.M. to 4:00 P.M.	40
4:00 P.M. to 8:00 P.M.	84
8:00 P.M. to 12 midnight	32

Determine the mean and median in units of cars per hour.

7.6. The rate of water flow to a processing plant varies with the time of day. The hourly rate data in gallons per minute (gpm) are 250, 200, 205, 290, 380, 500, 490, 400, 450, 500, 420, 450, 390, 380, 425, 505, 530, 500, 440, 400, 350, 300, 300, 290. Determine the mean, median, and mode.

7.7. A small machine shop has an inventory of galvanized metal sheeting. The numbers represent the size of each sheet: 8, 8, 9, 9, 9, 10, 10, 11, 12, 13, 14, 14, 14, 15, 16, 17, 17, 19. Determine the mean, median, and mode. How would you best represent the group with these centrality indices, bearing in mind that there are no fractional sizes?

7.8. Two sets of data have mean values of 30, but set A has a standard deviation of 2.0 and set B a standard deviation of 5.0. Sketch the differences between the two sets of data.

7.9. Using the data set from Problem 7.7 and letting the number of occurrences of a number be its weighting factor, determine the weighted average for the group.

7.10. The specific heat of a gas mixture is found by using the weighted average of the individual gas-specific heats. The following table lists the gases, weighting factors, and specific heats.

Gas	Weighting factor	Specific heat (kJ/kg-K)
Carbon dioxide	0.30	0.844
Sulfur dioxide	0.30	0.6225
Helium	0.20	5.1954
Nitrogen	0.20	1.0399

Determine the average specific heat of the mixture (the weighted average of the specific heats).

7.11. Determine the standard deviation for the data in Problems 7.1 and 7.2.

7.12. Determine the standard deviation for the data in Problem 7.3.

7.13. Determine the standard deviation for the data in Problem 7.6.

7.14. Write a computer program that will read data and calculate the arithmetic mean and the standard deviation.

7.15. Using the data in Problem 7.4, determine the standard deviation and plot the normal distribution curve.

7.16. In a manufacturing plant a steel rod is supposed to be cut to a length of 7.80 inches; however, measurements of the steel rods that were cut yielded the following lengths: 7.81, 7.80, 7.79, 7.80, 7.82, 7.80, 6.78,

7.80, 7.81, 7.83, 7.79, 6.77, 7.82, 7.80, 7.78, 7.80, 7.81, 7.81, 7.79, 7.87. Plot the frequency distribution plot, and determine the standard deviation. What is the maximum tolerance if you wish to assure that 95.45% of the rods are acceptable?

7.17. Bags of flour at a certain manufacturing facility are filled from a spring-loaded trap door that is opened for a certain time period. The bags of flour weigh, on average, 5 kg, but the standard deviation is 100 g. Considering that the plant produces 10,000 bags per day, determine the number of bags of flour weighing less than 5 kg; the number of bags of flour weighing more than 5.1 kg; and the number of bags of flour between 4.9 and 5.0 kg.

7.18. In a construction project many concrete samples are taken. For one particular project 3000 samples were taken, and they had an average compressive strength of 3500 psi. The standard deviation of the group was 250 psi. Determinne the number of samples with compressive strengths less than 3000 psi and the number of samples between 3250 psi and 3750 psi.

7.19. At the end of the first semester of the freshman year an engineering department found that the students had the following age distribution.

Age (years/months)	Number of persons
18/1–3	2
18/4–6	16
18/7–9	32
18/10–12	30
19/1–3	27
19/4–6	21
19/7–9	17
19/10–12	5
20/1–3	6
20/4–6	8
20/7–9	3
20/10–12	0

Plot a histogram of the data and determine the standard deviation. Can you assume a normal distribution for these data?

7.20. Determine the equation of the straight line that best represents the following (x, y) data pairs: 3, 5; 4, 4.50; 5, 6.50; 6.5, 7.0; 8.0, 8.0; 9.0, 9.50; 11.0, 12.0; 13.0, 11.50.

7.21. The temperature along a metal pipe is given by the following temperature and length data.

Temperature (°F)	Distance (in)
180	4.3
360	7.5
540	11.4
720	15.4
900	19.9

Determine the equation of a straight line representing these data.

7.22. The following data are from measuring the force and deflection data from a spring. Determine the equation of a straight line representing the data.

Force (lbf)	Deflection (in)
10	0.1
20	0.21
30	0.25
40	0.38
50	0.49
55	0.56

7.23. An orifice in a pipe has a coefficient determined by the equation

$$C = \frac{4W}{\pi D^2 t} \sqrt{\frac{1}{2g_c \rho \Delta p}},$$

where C is the coefficient, W is the weight, D is the diameter, t is the time, ρ is the density, Δp is the pressure drop, and g_c is a constant. The following information is known about the accuracy of the measurement: the weight to 1%, the diameter to 0.2%, the time to 1%, the density to 0.2%, and the pressure to 1%. Determine the percent error in the coefficient. Which term would you invest more money in to improve the accuracy of the coefficient?

7.24. In heat transfer the Nusselt number, Nu, is given by the following equation,

$$Nu = 0.023 \left(\frac{vD}{v}\right)^{0.8} (Pr)^{0.4},$$

where v is the velocity, D is the diameter, v is the kinematic viscosity, and Pr is the Prandtl number. To determine the Nusselt number the following data were obtained:

$D = 0.03$ m within 0.001 m

v = 50 m/s within 3 m/s

$v = 1.006 \times 10^{-6}$, a constant

$Pr = 7.0$ within 0.2 (dimensionless units)

Find the value of the Nusselt number and the percent error in its value.

CHAPTER 8

Concepts of Problem Solving

CHAPTER OBJECTIVES

To analyze a problem from its structure.

To present results in a professional manner.

To become familiar with the unit systems most often encountered in engineering practice.

To learn decision analysis techniques.

(Photo © Cameramann International at Computervision Corporation.)

Your course content in engineering may be thought of as having two components, analysis and design. Analysis is another name for problem solving, design another name for creativity. In this chapter you will learn about the analysis component of your engineering education.

8.1 PROBLEM FORMAT

Engineers solve problems. We are educated to solve them and like to solve them. A great deal of your education will be spent in solving a variety of problems, and you must learn to do this well and present the results in a fashion that is readily understood by the instructor and by you. Yes, you. You will use your homework problems when reviewing for tests and as references in later courses. You must include sufficient detail to make them useful to you when the material is not fresh in your mind.

Some fairly common formats are used to present homework and quiz and test results; your instructor will describe variations that she or he prefers.

Figure 8.1 illustrates one type of solution format. You may be asked to copy the problem statement, write what's given and what you are to find. All these formats have general similarities.

1. Alway show units, such as *lbf*, on your answers, unless the term has no units.
2. Show all your calculations clearly so you and your instructor know what you are doing.
3. When you have arrived at an answer, check its reasonableness. Is it the same order of magnitude as the other terms? Check your work, assumptions, and calculations.
4. Clearly identify your answer, underlined several times.
5. Do not start another problem on the same page unless it can be finished on that page.

Showing the units on the equations will help you in defining the problem; for instance, all the terms in a force balance should have units of force. Presenting the results in this somewhat formal fashion will require you, in many instances, to do a rough draft and then copy it over. Do not expect that the first attempt will always be presentable. The calculational procedure will help your instructor readily pinpoint a difficulty. If it is caused by a conceptual misunderstanding, you will be able to comprehend where the difficulty occurred in the context of the entire solution.

A misplaced decimal point can cause strange numbers to appear in part of the solution; see if the answer makes sense physically. Practice in problem

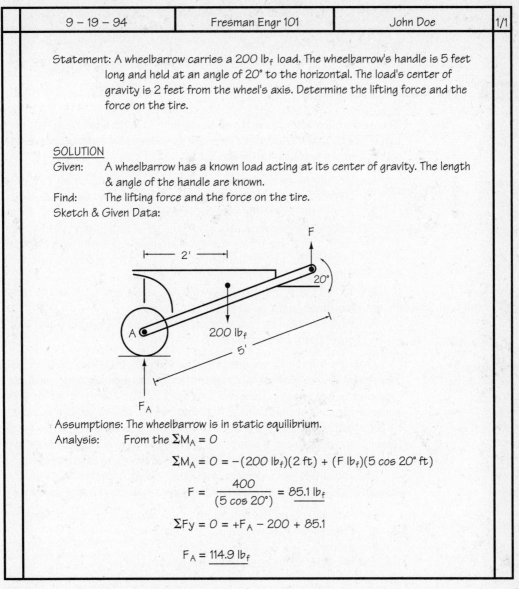

Figure 8.1 One type of solution format.

solving will also develop your engineering intuition. It is discouraging to have numerical results that seem wrong; you end up doubting your solution procedure when indeed only a calculation error occurred. Identify the answer for all to see. Remember you will be using your homework as a study guide later, so make it worthwhile. Minimizing the number of problems on a sheet also helps when you study the material later. Problems running over to another sheet require you to constantly turn between two pages, sometimes losing your continuity of thought. Very often students are cheap when it comes to using paper, and while it is laudable to conserve our natural resources, white areas around the sketch and numbers help tremendously in the problem's readability.

8.2 PROBLEM SOLVING

The assumption above is that you know how to solve a problem, but perhaps this is not the case. Guidelines in problem solving that have helped others are implicit in the solution in Figure 8.1.

1. Define the system under consideration—the material, circuit, or system that is to be analyzed: the wheelbarrow in Figure 8.1.
2. After you define the system, visualize what is happening. Drawing a sketch helps tremendously with the visualization process. In Figure 8.1 you are lifting the load, sharing a portion with the wheel.
3. Identify the given (known) material and the unknowns to be solved for.
4. Model the system with appropriate equations, such as the sum of the forces must be zero for equilibrium.
5. Perform the analysis necessary to solve for the unknowns (not necessarily an easy task).

Problem solving is not easily learned just by following steps. The ones mentioned above give us a direction to proceed in, but the path in that direction is not always clear. We can increase the clarity if we understand some of the conceptualizations that must occur before we can intelligently, rather than by rote, solve a variety of problems. Much of what you have learned about problem solving was developed in your high school algebra classes. Having a sound algebraic ability will help you in engineering.

The intuititve technique for problem solving has four steps that often heppen so quickly we don't realize they have occurred. Simply stated they are

1. Understanding the statement of the problem (a sketch helps),
2. Finding what is required,
3. Finding what facts are given, and
4. Deciding what operations are needed.

This is a more general rendition of the steps outlined previously. This listing stresses two important aspects that you might have missed earlier. From the first step you must understand the language that is used to describe the problem; you must be able to visualize (sketch) it if it is a physically real system. Part of your course material will help you to understand the language and its implications, and you will be able to visualize systems correctly. If you cannot, see your instructor. You will have pinpointed an area of confusion, so he or she can more easily be of assistance. You will not be successful in problem solving if this visualization and understanding are lacking.

Normally in textbook problems the information given (Step 3) and what is required (Step 2) are clearly identified. In some advanced areas this is not the case. Step 4, however, often lacks clarity. How do you decide what operations are required? Poor decisions here are most often the cause for failure. Once the equations are written, you and others can usually solve them.

Certain techniques will enable you to make better decisions. The key word is *connections,* the various connections between the different quantities in the problem statement. Let's see what this means by looking at an example.

EXAMPLE 8.1

The volume of a solid is 40 cubic centimeters. The solid is composed of two substances, one with a density of 6 g/cm^3 and the other with a density of 12 g/cm^3. The solid weighs 396 grams. Determine the volume that each substance occupies.

SOLUTION

Given: The volume and weight of a solid and the densities of the components that the solid is made from.

Find: The volume of the solid that each component occupies.

Sketch and Given Data:

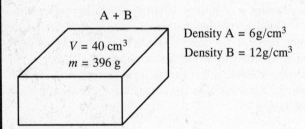

Figure 8.2

Assumptions: There is no loss of volume when the components are mixed.

Analysis: The first reading of the problem should bring into focus the differences among quantity, number and units. Density is a quantity, its numerical value is 6 for one substance and 12 for the other, and it has units of g/cm^3. Thus, we must understand philosophically what density is, what the units of it are in this particular situation, and what numerical value is associated with those units. With the first reading we look to establish a connection between the quantities that are given in the problem statement, Step 1. In this case there are four quantities involved: components in volume, mass per unit volume of component, total volume and total mass.

Next we establish the equations that relate to the connections between the quantities. If we multiply the component density by the component volume, we obtain the component mass. This, when added to the other component's mass, will give the total mass.

Total Mass = Sum of component masses

X = Volume of Component A $6X$ = Mass of A

$40 - X$ = Volume of Component B $12*(40 - X)$ = Mass of B

$(396 \text{ g}) = (6X \text{ g}) + (12*(40 - X)\text{g})$

$X = 14 \text{ cm}^3$ Volume of Component A

$40 - X = 26 \text{ cm}^3$ Volume of Component B

The reasoning techniques enable you to take the information given in the problem statement and manipulate it. In some instances reasoning, or processing the information, is not the difficulty; gathering the information creates the trouble. Figure 8.3 illustrates an information diagram of two general skills, processing information (reasoning) and gathering information. Reasoning ability involves inductive and deductive thinking and logic. Information gathering ability relates to reading comprehension, visualization, memory, and recall.

We would all like to be in quadrant I, have good abilities in reasoning and information gathering. However, in some courses you may find that you possess such abilities while in others you do not. The instructor may affect your attitude, hence ability, about the subject matter, making it easier or more difficult to assimilate. In other instances you need to pay special attention to developing your ability in either the gathering or processing areas.

Quadrant II denotes people who gather the information correctly, take good notes, understand what they read, and know what is asked of them, but have difficulty knowing how to proceed with the problem solution. For these people, following the formal problem solving procedures will be of assistance. Checking with the instructor about the logic of the solution is clearly important.

Quadrant II	**Quadrant I**
Good information gathering Poor information processing	Good information gathering Good information processing
Quadrant III	**Quadrant IV**
Poor information gathering Poor information processing	Poor information gathering Good information processing

Figure 8.3 An information diagram.

People in quadrant IV have the opposite problem. They can process the information correctly, once they know what the information is. Skills in note taking, reading comprehension, memorization, and visualization of what is happening are usually important in strengthening this skill. These people often need to focus on understanding technical words and the implications of certain terms.

People in quadrant III have to work hard to overcome deficiencies in two skill levels. Like those in quadrant IV, they must improve their information gathering skills, and then learn to process this information correctly, like those in quadrant II.

Both aspects of problem solution can be improved by being aware of what it is you do not understand. Is it the vocabulary? The visualization? Or does the difficulty lie in what to do with the information once you have it? By answering these questions you can focus your efforts on the area of weakness.

8.3 SIGNIFICANT FIGURES AND SCIENTIFIC NOTATION

The use of scientific notation for numbers is a great help when using large numerical values. For instance, if the energy used in a power plant is 185 million kilowatts and this number has to be used in calculations, it becomes cumbersome to write and manipulate. Expressed in powers of ten (scientific notation—see Table 8.1), the number is easier to work with: 1.85×10^8.

To use scientific notation accurately and correctly we need to understand the concept of significant figures. Suppose the result of a calculation is 7.68321, using a hand calculator. Are all those numbers after the decimal important, useless, or misleading? The numbers used in a calculation came from somewhere, resulted from measurements of some kind. The accuracy of

Table 8.1 SCIENTIFIC NOTATION

1 000 000	=	1×10^6
100 000	=	1×10^5
10 000	=	1×10^4
1 000	=	1×10^3
100	=	1×10^2
10	=	1×10^1
1	=	1×10^0
0.1	=	1×10^{-1}
0.01	=	1×10^{-2}
0.001	=	1×10^{-3}
0.0001	=	1×10^{-4}
0.00001	=	1×10^{-5}
0.000001	=	1×10^{-6}

the measurement determines the number of significant figures, the degree of correctness of the number. If you are measuring the thickness of an electric cable, you might use a ruler and find it is three-quarters of an inch in diameter. How sure are you of this measurement? As accurately as you can read the ruler, in this case 0.75 inches within a thirty-second of an inch, or ±0.03 inches. Thus, the answer would be 0.75, as that is the degree to which you are certain of the measurement. It would be erroneous to write 0.750, as that implies you know the value to the thousandth of an inch. Or you might use a micrometer and determine the measurement to be 0.740 ± 0.002; in the case your reading is accurate to the nearest thousandth of an inch, and the extra decimal place should be used to convey this information.

Number	Significant Figures
832	3
19	2
21.61	4
0.621	3
401.612	6
0.02	1
0.038	2
0.02000	4

Returning to scientific notation, the steps involved for expressing a number are

1. Locate the decimal point so the number has a value between one and ten.
2. Use only the number of digits that are significant figures.
3. Multiply the number by powers of ten to obtain the correct magnitude.

The value of 986 in scientific notation is 9.86×10^2; the value of 986.00 is 9.8600×10^2 because the decimal point in the original figure indicated the greater accuracy of the number, five significant figures versus three for 986. If the number is given as 98,600, we cannot tell whether the number should be written as 9.86×10^4, 9.860×10^4, or 9.8600×10^4. The initial number does not convey its significant figures accurately. In textbooks, an immediate concern of yours, consider all numbers to have the maximum significant figures possible unless you are given information to the contrary.

The value of scientific notation is that it allows you to inform others as to the accuracy of the number, its significant figures. In addition it allows you to handily express large numbers used in calculations. Hence, $5,630,000 \times 0.00000609 = 34.3$ becomes $(5.63 \times 10^6) \times (6.09 \times 10^{-6}) = 3.43 \times 10^1$.

For addition, subtraction, multiplication, and division of numbers with varying significant figures, the result is only as accurate—has as many significant figures—as the number with the least significant figures. Thus, when we add the following figures, $28.631 + 4928.1 + 0.9763 = 4957.7073$, the second term is accurate to only the tenths place, so the correct answer is 4957.7. Multiplication and division follow the same rule. Thus, a calculator might display the following result, 27.3 × 0.93651 = 25.566723, but the corrected answer is 25.6. The other numbers are rounded off.

The rule for rounding off is that if the digit to the right of the last significant figure is five or greater, round the last significant figure up; if it is less than five, round the figure down. If 31.6952 and 34.6949 are rounded off to four significant figures, the results are 31.70 and 34.69. Conversion factors (2.54 centimeters per inch) or integer values used when performing arithmetic operations are considered to contain unlimited significant figures. The other terms dictate the number of significant figures in the answer.

8.4 UNIT SYSTEMS

The numbers engineers work with in most instances have dimensions associated with them. We use these numbers to characterize systems by their length, mass, temperature, and other standards. When these standards are combined into a coherent set, they form a measurement system. The English engineering system and the SI metric system are the two systems most frequently encountered in engineering practice and in college texts. Other systems are used; however, if you understand the bases for these two, others are similar in terms of their development.

FUNDAMENTAL AND DERIVED UNITS

Units that are postulated, or defined, are called fundamental units. Once a few units are defined, others can be derived from them. Some of the fundamental units are length, time, mass, force (in the English system), and temperature.

Length (L) is the distance between two points in space.

Time (t) is the period between two events or during which something happens.

Mass (m) is the quantity of matter that a substance is composed of. It is invariant with location: the mass of a person on earth or on the moon is the same.

Force (F) is defined through Newton's second law; conceptually it is associated with power or strength, the push or pull on an object. It may occur directly, such as the push on a wheelbarrow, or indirectly by fields (gravitational, electrical, magnetic) acting on an object. Force is sometimes viewed as a fundamental unit, as it is in the English engineering system, and sometimes as a derived unit, as in the SI system.

Temperature (T) measures the degree of hotness or coldness of an object. We commonly use arbitrary temperature scales, such as the Fahrenheit and Celsius scales. Temperature scales are for the most part based on two points, the triple point of water (where ice, liquid, and vapor coexist) and the boiling point of water at atmospheric pressure. The Fahrenheit scale was named for Gabriel Daniel Fahrenheit. He was interested in thermometry, as was an astronomer friend, Romer, who devised a temperature scale of 60 degrees.

There are 60 seconds to a minute, 60 minutes in an hour, so why not 60 degrees for a temperature scale? Fahrenheit thought the scale too rough, and increased the number of divisions to 240; why not 360 divisions remains a mystery. The lower limit was an ice, salt, and water mixture that corresponds to 0°F, and the human body temperature was set at 90°, which Fahrenheit later decided should be 96°. With the scale fixed by body temperature and an ice, salt, and water mixture, the triple point became 32° and water boiled at 212°. To eliminate these inconvenient numbers a Swedish astronomer by the name of Anders Celsius devised a scale that started at 100° (triple point) and went to 0° (boiling point). A friend suggested he reverse them, which resulted in the current Celsius temperature scale. Lord Kelvin developed the idea, championed by others earlier, of the absolute temperature scale and devised the Kelvin temperature scale, which uses the divisions of the Celsius scale. Another absolute temperature scale is the Rankine scale, which uses the divisions of the Fahrenheit scale. The relationships between the scales are

$$F = 9/5 \; C + 32$$
$$C = 5/9 \; (F - 32)$$
$$K = C + 273.16$$
$$R = F + 459.67$$

where F is degrees Fahrenheit, C degrees Celsius, K degrees Kelvin, and R degrees Rankine. Figure 8.4 graphs the four temperature scales.

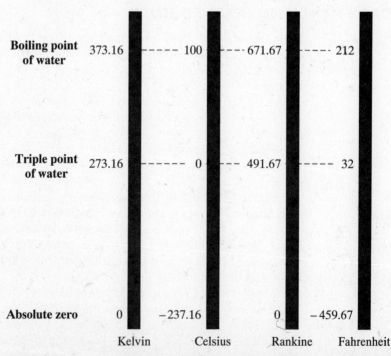

Figure 8.4 The graph of four temperature scales.

Other units can be derived from fundamental units. Area and volume are derived from length, L^2 and L^3. Velocity is distance per unit time, or L/t, and pressure is force per unit area, or F/L^2. Unit systems assign numerical values to these dimensions, and we normally use these dimensions in conjunction with the unit systems, merging the two concepts in our minds.

ENGLISH ENGINEERING UNIT SYSTEM

The English engineering unit system developed in a rather haphazard fashion, with many common measures, such as the yard, descended from such standards as the length from the thumb to the nose of Henry I of England. Clearly, such a system has many drawbacks, but a singular advantage is that we know it well, having grown up with feet and pounds.

One drawback is the confusion that results from having force and mass as fundamental units. They are related through Newton's second law, which states that the force acting on a body is equal to the mass times acceleration.

$$F = ma, \qquad 8.1$$

where F is measured in pounds force (lbf), m is measured in pounds mass (lbm), and a is measured in feet per second squared (ft/sec^2). One pound force is defined as one pound mass accelerated 32.174 ft/sec^2. When we substitute the values of the definition into Equation 8.1, a problem appears in the unit balance of the equation: lbf = (lbm)(ft/sec^2). The units on one side do not equal the units on the other side, so let's multiply Equation 8.1 by a constant that has units that balance the equation:

$$F = kma, \qquad 8.2$$

or with the units substituted,

$$\text{lbf} = k(\text{lbm})(\text{ft/sec}^2).$$

Thus, the units of the constant must be lbf-sec^2/lbm-ft. To determine the value of the constant, we substitute into Equation 8.2 the definition of one pound force and solve for k:

$$k = \frac{1}{32.174} \times \frac{\text{lbf-sec}^2}{\text{lbm-ft}}. \qquad 8.3$$

Since Equation 8.2 is valid for all cases, if k is determined for a given case, it must be true for all the other cases. Standard gravitational acceleration, g, is 32.174 ft/sec^2, a term with the same numeric value as k but different units. The term k was noted to be equal to $1/g_c$, where g_c = 32.174 lbm-ft/lbf-sec^2, and thus Equation 8.2 becomes

$$F = ma/g_c \text{ lbf.} \qquad 8.4$$

EXAMPLE 8.2

Consider that a person seated in a car is uniformly accelerated at 5 ft/sec² and that the person has a mass of 180 lbm. Find the horizontal force acting on the person.

SOLUTION

Given: A person is accelerated horizontally.
Find: The horizontal force as a result of this acceleration.
Sketch and Given Data:

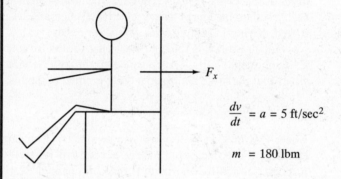

$\frac{dv}{dt} = a = 5 \text{ ft/sec}^2$

$m = 180 \text{ lbm}$

Figure 8.5

Assumptions: The acceleration is uniform and constant.
Analysis: Substitute into Equation (8.4) to determine the horizontal force.

$$F = ma/g_c = \frac{(180 \text{ lbm})(5 \text{ ft/sec}^2)}{32.174 \frac{\text{lbm-ft}}{\text{lbf-sec}^2}} = 27.97 \text{ lbf}$$

Comments: Force acts in the direction of the applied acceleration or deceleration. Thus, if the acceleration acts in more than one dimension, there will be force components in those directions, with the resultant force being a vector in the same direction as the acceleration vector.

Thus far we have not mentioned weight. Take the man in Example 8.2 and suppose he is standing on the ground. He experiences a force on him equal and opposite to the force he exerts on the earth. This force is his weight. He is acted upon by the local gravitational field g. If the gravitational field is located so g has a value of 32.174 ft/sec², the standard gravitational acceleration, then the force is

$$F = \frac{(180 \text{ lbm})(32.174 \text{ ft/sec}^2)}{32.174 \text{ (lbm-ft)/(lbf-sec}^2)} = 180 \text{ lbf.}$$

Table 8.2 UNITS IN THE ENGLISH ENGINEERING SYSTEM

Fundamental		Derived	
Force (F)	lbf	Area (L^2)	ft^2
Mass (M)	lbm	Volume (L^3)	ft^3
Length (L)	ft	Acceleration (L/t^2)	ft/sec^2
Time (t)	sec	Density (M/L^3)	lbm/ft^3
		Pressure (F/L^2)	lbf/ft^2
		Energy (FL)	ft-lbf
		Power (FL/t)	(ft-lbf)/sec

If the man were standing on a tall mountain where g has a value of 30.0 ft/sec^2, then the force exerted on the ground, his weight, would be

$$F = \frac{(180 \text{ lbm})(30.0 \text{ ft/sec}^2)}{32.174 \text{ (lbm-ft)/(lbf-sec}^2)} = 167.8 \text{ lbf}.$$

In space, the condition of weightlessness is caused by the fact that $g = 0$; hence the force or weight of the body mass is zero.

Table 8.2 lists the fundamental and some of the derived units found in the English engineering system.

SI UNITS

SI stands for Système International d'Unités, a consistent unit system developed in 1960. In 1975 the metric conversion act became law, declaring in part "a national policy of coordinating the increasing use of the metric system in the United States." The process of conversion has been slow, but as the business outlook of the United States becomes more global, the United States will convert of necessity to metric sizes. It is easier to perform calculations in metric than in English engineering units. The difficulty in the conversion process is that standard sizes of equipment, pipes, bolts, and parts of all kinds must be changed. This requires a major investment in new manufacturing tools, which businesses may not be able to afford, but eventually they must convert if they are to compete in the world marketplace where metric is standard.

SI has seven physically defined units—fundamental or base units—and two geometrically defined supplementary ones, and the rest are derived. The fundamental and supplementary units are listed in Table 8.3. Each of these terms is rigorously defined.

A meter is the distance in space that is equal to 1.650 763 73 $\times$ 10^6 wavelengths in vacuum of the radiation corresponding to the transition be-

Table 8.3 FUNDAMENTAL UNITS IN SI

Quantity	Name	Symbol
Base units		
Length	meter	m
Mass	kilogram	kg
Time	second	s
Electric current	ampere	A
Thermodynamic temperature	kelvin	K
Amount of substance	mole	mol
Luminous intensity	candela	cd
Supplementary units		
Plane angle	radian	rad
Solid angle	steradian	sr

tween two energy levels of the krypton 86 atom. This is not a handy definition, but is certainly a precise one. For everyday conceptualization a meter is 39.37 inches.

A kilogram is a mass equal to the mass of an international prototype made of a platinum-iridium alloy and kept at the International Bureau of Weights and Measures in France. Relating this to everyday life, it is equal to 2.205 pounds mass.

A second is defined as the duration of $9.192\ 631\ 770 \times 10^9$ periods of radiation in the transition between the two hyperfine levels of the ground state of the cesium 133 atom. Originally a second was based on a fraction of the mean solar day, which used the earth's rotation as a clock, and this is the basis of our day-to-day observation of time.

An ampere is the constant current that, if maintained in two straight parallel conductors of infinite length and of negligible circular cross-section and placed one meter apart in vacuum, would produce between these conductors a force equal to 2×10^{-7} newtons per meter of length. Again in daily life this is essentially a flow of one coulomb per second, where one coulomb is equal to 3×10^9 units of charge.

A kelvin degree is 1/273.16 of the thermodynamic temperature of the triple point of water.

A mole is the amount of substance in a system that contains as many elemental entities as there are atoms in 0.012 kilograms of carbon 12.

A candela is the luminous intensity, in the perpendicular direction, of a surface of 1/600 000 square meter of a blackbody at the temperature of freezing platinum under a pressure of 101 325 newtons per square meter.

A radian is a unit of measure of a plane angle with its vertex at the center of a circle and subtended by an arc equal in length to the radius.

Table 8.4 SI UNIT PREFIXES

Factor by Which Unit is Multiplied	Prefix	Symbol
10^{-18}	atto	a
10^{-15}	femto	f
10^{-12}	pico	p
10^{-9}	nano	n
10^{-6}	micro	μ
10^{-3}	milli	m
10^{-2}	centi[a]	c
10^{-1}	deci[a]	d
10^{1}	deka[a]	da
10^{2}	hecto[a]	h
10^{3}	kilo	k
10^{6}	mega	M
10^{9}	giga	G
10^{12}	tera	T
10^{15}	peta	P
10^{18}	exa	E

[a] Avoid these prefixes, except for centimeter.

A steradian is a unit of measure of a solid angle with its vertex at the center of a sphere and enclosing an area of the spherical surface equal to that of a square with sides equal in length to the radius.

Force is a derived unit in SI; one unit of force is defined by Newton's second law, Equation 8.1. A unit force is a newton (N), so 1 N = (1 kg)(1 m/s^2). The potential for confusion between mass and force does not exist, as they not only have different names but different numerical values as well. For instance, the force exerted on a 40 kilogram woman by standard gravitational acceleration, 9.8 m/s^2, is

$$F = (40 \text{ kg})(9.8 \text{ m/s}^2) = 392 \text{ N}.$$

Table 8.4 shows the prefixes that indicate the order of magnitude of the term and derived units in SI. Table 8.5 lists the derived units in SI.

SI symbols are strictly interpreted. SI associates an exact meaning to each symbol, and one of its values lies in this preciseness. On the other hand, it does not allow for variations from these rules, as these variations would imply some other meaning. For instance in English units we may write ft-lbf or lbf-ft and mean the same thing, though the first expression is desirable. In SI N m and mN are entirely different: the former is newton meter and the latter is millinewton. Notice that a space was left between N and m in the first term;

Table 8.5 SI DERIVED UNITS

Quantity	Unit	SI Symbol	Formula
Acceleration	meter per second squared	—	m/s^2
Angular acceleration	radian per second squared	—	rad/s^2
Angular velocity	radian per second	—	rad/s
Area	square meter	—	m^2
Density	kilogram per cubic meter	—	kg/m^3
Electric capacitance	farad	F	A · s/V
Electrical conductance	siemens	S	A/V
Electric field strength	volt per meter	—	V/m
Electric inductance	henry	H	V · s/A
Electric potential difference	volt	V	W/A
Electric resistance	ohm	Ω	V/A
Energy	joule	J	N · m
Entropy	joule per kelvin	—	J/K
Force	newton	N	kg · m/s^2
Frequency	hertz	Hz	l/s
Illuminance	lux	lx	lm/m^2
Luminance	candela per square meter	—	cd/m^2
Luminous flux	lumen	lm	cd · sr
Magnetic field strength	ampere per meter	—	A/m
Magnetic flux	weber	Wb	V · s
Magnetic flux density	tesla	T	Wb/m^2
Power	watt	W	J/s
Pressure	pascal	Pa	N/m^2
Quantity of electricity	coulomb	C	A · s
Quantity of heat	joule	J	N · m
Radiant intensity	watt per steradian	—	W/sr
Specific heat	joule per kilogram-kelvin	—	J/kg · K
Stress	pascal	Pa	N/m^2
Thermal conductivity	watt per meter-kelvin	—	W/m · K
Velocity	meter per second	—	m/s
Viscosity, dynamic	pascal-second	—	Pa · s
Viscosity, kinematic	square meter per second	—	m^2/s
Volume	cubic meter	—	m^3
Work	joule	J	N · m

this is the correct way to write the combination of SI units. The following is a list of some important rules for using SI.

1. Unit symbols are printed in lowercase roman letters. Periods are not used after a symbol except at the end of a sentence. Thus, we would write N or kg, and not N. or KG. Italic letters are used for quantity sumbols; thus, m is mass and m is meter. Additionally the symbol is the same in the singular and plural: thus, 1 kg, 10 kg.
2. The prefix precedes the symbol in roman type without a space, and double prefixes are not allowed: Thus, mA and not μmA.
3. For distance in engineering and architectural drawings, use millimeter as the basic unit.
4. The product of two or more units is represented by a space or a dot between the units; thus, N m or N $\cdot$ m.
5. Where the division of terms must occur the slash mark should not be used more than once, nor should a hyphen be repeated on the same line, as ambiguity results in both cases. In the case of acceleration, m/s^2 or m-s^{-2} are acceptable, but m/s/s or m-s-s is vague. It could be interpreted that the seconds cancel one another and meters remain.
6. Capital and lowercase letters must be used as described; for instance, K means degrees Kelvin and k means kilo.
7. Don't use commas in long number. In some countries commas are used in place of decimal points. Rather, use a space after every third digit to the left and right of the decimal point. If there are only four digits the use of the space is optional. Thus, you would write 3 600 000 and 21.005 68, not 3,600,000 or 21.00568.
8. When writing a number less than one, always put a zero before the decimal point; 0.625, not .625.

Some of these rules are difficult to use when you are writing by hand, particularly when differentiating between roman and italic letters; adopt a different symbol for clarity, such as a script m for mass, so your notes are not confusing.

8.5 CONVERSION FACTORS

It will remain necessary to convert between unit systems for many years, as different unit systems are in use around the world. A conversion factor helps us convert from the value in one system to its value in another. Appendix 2 gives a list of conversion factors, and we will shortly demonstrate their use in several situations. But more fundamentally we should understand why they can be used: they are dimensionless ones, and hence we can multiply or divide any equation by them and not change the value of the equation. Let's examine the conversion between inches and centimeters, 1 inch = 2.54 centimeters. Divide this equation by 1 inch, yielding 1 = 2.54 cm/in. Thus, 2.54 cm/in is equal to a dimensionless one.

EXAMPLE 8.3

A cargo ship has large tanks within it for carrying fuel oil. The oil is meaured in barrels, while the tank dimensions are in meters, 1 m × 5 m × 15 m. How many barrels does the tank hold? How many gallons?

SOLUTION

Given: The size of a ship's tank in meters.

Find: The number of barrels and gallons that the tank volume holds.

Sketch and Given Data:

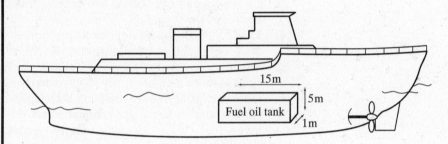

Figure 8.6

Assumptions: None.

Analysis: Calculate the volume in cubic feet, then find in the conversion tables the value to convert from cubic meters to barrels.

$$V = 1 \times 5 \times 15 = 75 \text{ m}^3$$

The conversion table does not have a direct conversion factor, so we must convert from cubic meters to gallons and from gallons to barrels.

$$V = (75 \text{ m}^3)(2.642 \times 10^2 \text{ gal/m}^3) = 19\,815 \text{ gal}$$

Again we must calculate a new conversion factor, as we are given that 42 gal equals 1 bbl. The conversion factor is 1/42 bbl/gal and the volume in barrels is

$$V = (19\,815 \text{ gal})/(42 \text{ gal/bbl}) = 471.8 \text{ bbl}$$

Comments: Often a direct conversion factor is not available, and we must perform intermediate steps.

8.6 TECHNICAL DECISION ANALYSIS

Engineers at all levels, from entry level to senior management, need to make decisions, assessing the various factors underlying the decision and then reaching a logical conclusion about it. As an engineering manager you might have to decide whether or not to invest in a new product line with its related machinery costs, or use the money elsewhere. There will be no clear-cut yes-or-no answer: uncertainty is attached to whatever choice you make. Techniques that assist in decision making will be investigated here. As a manager you will be judged by the decisions you make and, unlike a baseball batting average, judged by the last few decisions and whether or not they were correct, not by the total number of correct decisions. You must make decisions in situations where you will have insufficient information or imperfect information. You no longer need to shoot from the hip, however, as methodologies exist to minimize risk to any decision and maximize return.

DECISION TREE

Your company owns land on which there may or may not be oil. A competitor of yours is willing to lease the land from you for $500,000, and you must make a decision within ten days. If you accept the lease, the company receives $500,000. If you reject the proposal, you have two choices, to drill for oil or not. If you decide to drill for oil there are four possible outcomes: natural gas, no oil, occurring 40 percent of the time; natural gas and oil, occurring 30 percent of the time; dry hole, occurring 20 percent of the time; and oil, occurring 10 percent of the time. The cost of drilling is $1,500,000, and from past experience you know that the value gained by the company is zero for the dry hole; $1,500,000 for natural gas only; $3,000,000 for gas and oil mixture; and $6,000,000 for oil only. What decision should you make, assuming that incurring a loss of $1,500,000 will not bankrupt the company?

To help answer this question you might draw a decision tree, as shown in Figure 8.7. This tree represents all the possible decisions, their outcomes, and the probability of each outcome. Initially there are two choices, to accept the

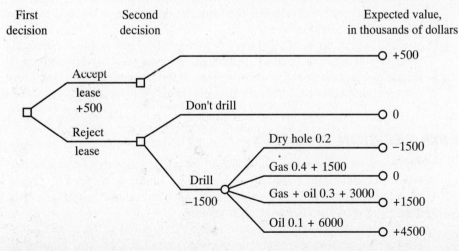

Figure 8.7 A decision tree.

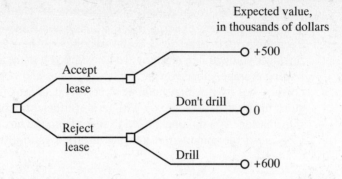

Figure 8.8 A simplified version of the decision tree in Figure 8.7.

lease or not to do so. If the lease is rejected, again there are two choices, to drill or not to drill. If you drill there are four possible outcomes; the net dollar value is listed at the far righthand side of the tree. How then to decide? The various options in the "to drill" branch may be combined into an average expectation by finding the weighted probability and the expected value (EV) associated with it.

$$EV = 0.2(0 - 1{,}500{,}000) + 0.4(1{,}500{,}000 - 1{,}500{,}000)$$
$$ + 0.3(3{,}000{,}000 - 1{,}500{,}000) + 0.1(6{,}000{,}000 - 1{,}500{,}000)$$
$$EV = +600{,}000$$

Thus, the expected value from drilling, considering all the various possibilities and their probabilities, is $600,000. You would choose then not to lease. The simplified decision tree is shown in Figure 8.8. You could still come up with a dry hole and lose $1,500,000, but perhaps there is a way to minimize this possibility by obtaining more information about the land by test drilling. This information has a cost, and it makes for a much more complicated decision tree, but the value of the information can be quantified to some degree by this analysis.

The area of decision analysis is very interesting and useful in estimating the value of a given decision. Although the example is oversimplified, it should give you an idea of what is involved. Industrial engineering, operations research, and business administration deal with problems of this type.

MULTICRITERIA DECISION ANALYSIS

The previous example had one criterion for judging the merits of the various courses of action, the expected value. Often a situation has several criteria, of various degrees of importance, which must be evaluated simultaneously. This requires multicriteria decision analysis.

Consider the situation where you have job offers from two totally different companies. How might you decide which position to accept? First establish the criteria you will use in making a decision. Then determine the relative importance, the weighting factor, associated with each one. Table 8.6 lists the

Table 8.6 CRITERIA AND WEIGHTING FACTORS IN JOB SELECTION

Criteria	Weight (1–10)
1) Salary and benefits	9
2) Location	7
3) Cost of living	5
4) Opportunity for advancement	7
5) Co-worker compatibility	5

Table 8.7 RESULTS OF MULTICRITERIA EVALUATION

Company A (1–10)	Company B (1–10)
1) $10 \times 9 = 90$	$8 \times 9 = 72$
2) $5 \times 7 = 35$	$7 \times 7 = 49$
3) $6 \times 5 = 30$	$6 \times 5 = 30$
4) $6 \times 7 = 42$	$8 \times 7 = 56$
5) $7 \times 5 = 35$	$8 \times 5 = 40$
Total 232	247

criteria and their weights. Determining the criteria and the weights is, in general, not an easy task if skewing the decision is to be avoided. For this example five criteria have been selected: salary and benefits are very important to you and are given a weight of one; location of the company—what part of the country, nearness to your family—is also important and is given a weight of seven; the cost of living affects how much of your income you will be left with after essentials are taken care of and is given a weight of five; important to your career development is the opportunity for advancement within the organization and is given a weight of seven; certainly how you relate to your co-workers affects your decision in selecting the company, hence the weighting of five.

You receive job offers from companies A and B and the salary for A is 15 percent better than that offered by B. The location is not as acceptable as B, though not disastrous; the cost of living is about the same in both locations; the route of advancement at A is more regimented than at B; and you liked the people you would be working with at B slightly more than at A. Table 8.7 shows the final analysis in rating the companies. You determine the relative grade each company received for each criterion, multiply by the weighting factor, and total. In this case company B is the choice.

GANTT CHART

Managing a project with interdependent operations, such as the construction of a building, requires the scheduling and controlling of various trades and material flows. From such a schedule, situations that may cause bottlenecks can be ascertained and extra supervision provided at these junctions. The simplest form of schedule is the Gantt chart, shown in Figure 8.9. As work is completed you color in the areas, allowing you to see the timeliness of the project. In Figure 8.9 the work is slightly ahead of schedule.

This method is useful for small projects but would become too cumbersome for those with many tasks. Nor does it indicate which parts of the schedule can fall behind and not hinder the progress of the project and which parts are critical. The methods that take these factors into consideration, such as critical path analysis, are beyond the scope of this text.

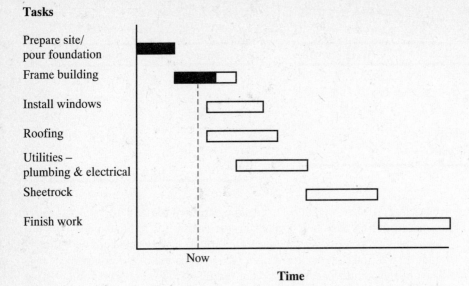

Figure 8.9 A Gantt chart of the schedule for the construction of a building.

REFERENCE

1. Yeshurum, Sharaga. *The Cognitive Method.* National Council of Teachers of Mathematics, Reston, VA, 1979.

PROBLEMS

8.1. As president of an engineering society at your school you have decided to celebrate winning the regional design contest. Fifty people are being invited; those you know will receive an invitation postcard, costing $0.18 in postage each, and those you do not know will receive a letter, costing $0.25 in postage. If the total postage paid is $11.10, determine the number of letters and postcards sent.

8.2. Determine the significant figures in each of the following numbers.
  a) 59.21 **b)** 642.01 **c)** 1.7002 **d)** 3.8
  e) 4.31×10^2 **f)** 8.3201×10^{-2} **g)** 0.009 00 **h)** 9100

8.3. Write the following numbers in scientific notation.
  a) 561.92 **b)** 8.31 **c)** 0.0396 **d)** 3101.25
  e) 729 034 **f)** 93 000 **g)** 0.0004 **h)** 43.10

8.4. Convert the following numbers from scientific notation to decimal notation.
  a) 8.341×10^2 **b)** $1.693\,21 \times 10^{-3}$ **c)** 9.8316×10^{-1}
  d) 9.3×10^4 **e)** 7.029×10^3 **f)** 3.0×10^{-2}

8.5. Perform the following additions, giving the answer to the correct number of significant figures.

a) 3.892
  0.2305
  25.9

b) 0.0695
  3.4509
  20.021

c) 4.50×10^3
  1.450×10^3
  2.31×10^2

d) 0.854
  0.0094
  0.021

e) 7.3×10^{-4}
  3.65×10^{-1}
  5.9×10^{-3}

8.6. Perform the following multiplications and divisions, giving the answer to the correct number of significant figures.

a) $(0.289)(0.067) =$
b) $(15.79)(0.5731) =$
c) $(90.620)(-12.21) =$
d) $(8957.1)(0.541) =$
e) $(6.893 \times 10^{-2})/(0.035) =$
f) $(7.346 \times 10^{-3})/(6.456 \times 10^2) =$

8.7. Convert the following.
a) 75°F to C
b) −40°C to F
c) 132°F to K
d) 10°R to K
e) 10°R to C
f) 50°C to F
g) 50°C to R

8.8. Convert the following.
a) 50 million gal/day to m³/s
b) 500 m/min to miles/h
c) 2000 lbm to kg
d) 20 gal/min to l/s
e) 88 ft/sec to km/h

8.9. An electric utility charges its customers 12 cents per kilowatt hour (kwh) of usage up to 2000 kwh and 15 cents per kwh above 2000 kwh. A business uses 25 kW per day for 20 days per month and 5 kW for the remaining 10 days. What is the monthly electric bill?

8.10. An electric power generation station uses ten tons of coal per hour. If the ash content of the coal is 12%, how many pounds of ash are produced each day?

8.11. In the equation $\dot{m} = \rho A v$, $\dot{m}$ is the mass flowrate, ρ is the density, A is the area, and v is the velocity. Find $\dot{m}$ in lbm/min if ρ is 60.0 lbm/ft³, A is 288 in², and v is 2 ft/sec.

8.12. In the equation $F = ma$, F is 100 N and m is 5 kg. Find the acceleration in m/s² and ft/sec².

8.13. The time required for a pendulum to complete one complete swing is $t = 2\pi(L/g)^{0.5}$, where L is pendulum length and g is gravitational acceleration. If the time is measured at two seconds and the length is one meter, determine the value of g in m/s². Convert the length to feet and compute the gravitational acceleration in English units (you may need g_c in the equation; perform a unit balance to check).

8.14. You have been appointed chair of a student committee that must solicit opinions from students, faculty, administrators, alumni, and corporate executives regarding a major engineering project. The time frame for the committee's final report is six months. During this time you must organize the committee, hold committee meetings, solicit opinions from the various constituencies, consolidate information, and write the final report. Determine the activities necessary to complete this project, and construct a Gantt chart for them.

8.15. Using the multicriteria methodology, develop a model for evaluating the following.
  a) the purchase of a new automobile
  b) the decision to live on or off campus
  c) changing universities
  d) changing from an engineering major to a nonengineering one

8.16. Repeat the decision analysis for the oil company in the situation where the competitor offers $400,000, the cost of drilling is $1,400,000 and the probabilities are 25% for a dry hole, 35% for gas only, 25% for gas and oil, and 15% for oil. Determine the decision tree for the analysis and what your recommendation would be and why.

CHAPTER 9

Engineering Design

CHAPTER OBJECTIVES

To study the design process, attuning yourself to the steps involved.

To initiate your own preliminary designs.

(Photo Courtesy of NASA.)

In this chapter you will examine and practice the steps in the engineering design process, directing your creativity towards the solution of design challenges.

In Chapter 3 one of the various careers in engineering that we investigated was design engineering. You will find that design forms a vital part of your education in engineering, no matter what your special field. The engineer transforms technical knowledge through inspiration, perspiration, and creativity into new products and systems that satisfy a need, solve a problem. This is engineering design. All fields of engineering require engineers to display this ability: constructing new buildings and roadways in civil engineering; building advanced rockets and aircraft in aerospace engineering; and devising new microchips and computer systems in electrical engineering.

When you seek and obtain your first job, companies do not expect you to know the leading edge of development in a given field. They can educate you quickly to this level. They cannot, however, teach design and rely on your educational program to develop your ability in this area. Hence, you need the design courses and design aspects in many of the courses you take as juniors and seniors.

9.1 THE DESIGN PROCESS

The design process is composed of several steps: stating the problem, creating possible solutions, refining and analyzing them, evaluating them and implementing the design solution.

Textbook problems are usually nicely defined, given this, find that. In the engineering world, however, such clean definition is not always possible, and if the definition is determined, finding what is known or what can be assumed is often perplexing. On large projects there can be underlying questions concerning time constraints on implementation; some solutions won't work because they are too time intensive. When dealing with a subsystem or a component within a total system, defining the boundaries is of critical importance. There are compatibility problems to consider; specifications must state the input and output requirements of the unit. These become the criteria by which the design is judged in later steps.

Consider that you are called upon to design the coal loading system for railroad cars. One underlying constraint affecting the design could be that the train must keep moving at a certain velocity during the loading process. Problems have constraints on them, parameters within which solutions must function.

What are the sources of engineering problems? There are many ways these problems arise. An observable deficiency in an existing product or

Problem definition was no easy task when considering landing an astronaut on the moon's surface. A variety of factors, some of which could only be guessed at, had to be considered before solutions were created. (Courtesy of NASA)

process creates many problems. Perhaps the state of the art may have changed and a new design is needed to meet new and typically higher expectations. Maybe the sheet feeder on a printer or copier jams with paper too easily or there is overheating of a computer chip. Another source of problems is that devices and processes must be created to fulfill unmet needs, needs created by development such as an irrigation system or a department store heating and air conditioning system. Product enhancement is another likely source. Once a product is manufactured it frequently needs to be improved, especially when following the TQM philosophy. Thus, a furnace may be redesigned to improve efficiency, a printer to be quieter and faster, a milling machine to achieve more accurate tolerances. Laws change as well, affecting building codes, the environment and other regulations. Products must be redesigned to conform with the new regulatory environment. For instance, automobiles are now designed to achieve greater fuel efficiency because of national laws. Many states do not allow the venting of gasoline vapor when filling the tank of an automobile so the hose dispensing connection had to be redesigned.

9.2 PROBLEM DEFINITION

The purpose of problem definition is to clarify the problem as much as possible. Again, engineers need to be effective communicators in ascertaining what the problem is. A problem may be to design a new lamp. What type of light—incandescent or fluorescent; for what price range; to sit on a table, be hung from the ceiling or attached to a wall; interior or exterior? These and other considerations need to be addressed by the engineer before starting the basic design. The engineer begins with a statement of need. The statement of need is somewhat general so as not to limit creativity by directing thoughts to a preconceived solution.

Consider the exercise bicycle manufacturer discussed in Chapter 4. There are several steps in the assembly process and materials need to be moved from one area to another. One way to state the problem of moving the parts is "design a dolly to move the parts from welding to machining." This statement of need already contains the solution, precluding other choices. A wiser statement would be "design a means to move parts from welding to machining." In this situation many possibilities may be examined including the dolly solution. Immediate investigation may yield basic requirements such that one person should be able to move the parts, the time for transfer should be less than five minutes and the means of transport must not interfere with the work or safety of others in the area.

There will be basic limitations that emerge with investigation, such as those imposed by laws or contracts. Regarding the transport of bicycle parts, the union contract does not allow any person to lift more than 50 pounds; the parts should not bump one another; there is a doorway that is four feet wide and eight feet high between the areas; and your boss requires a solution in two weeks' time.

Not every item of information is a limitation or a requirement, some is additional data. For instance, the parts will be in a variety of sizes ranging from 5 inches to 2 feet and weighing between 1 and 18 pounds; the number of parts vary between 10 and 50; and the distance to move the parts is 60 feet.

Frequently when a new product or process is to be designed, questions arise for which the engineer will seek answers. What floor space is available for loading the bicycle parts; how frequently does this happen; can the doorway be widened? Answers to these questions, when combined with the requirements, limitations, and data direct the engineering creativity.

9.3 CREATING SOLUTIONS

This phase of the design process is creating solutions for the problem. Notice that solutions, the plural, is used; there will be several ways to resolve the problem, some better than others, but they all should be examined initially. Do not prejudge or censor ideas before allowing them to surface. This is where your technical education and engineering experience combine with your innate creative ability to develop problem solutions. The ways that have been used in the past may be just fine for the present situation: no new technology or manufacturing methods are needed. However, inventiveness is required when research and development have created new technologies that can be used for new products and projects. For instance, the advances in current aircraft design could not have occurred without first developing the graphite epoxy materials that withstand greater forces than does high-strength aluminum, the material previously in use.

Often in the design process when solutions are first created, questions arise about the problem statement whose resolution better defines the statement. The solutions remain qualitative at this point.

Just as using a problem solving format (Chapter 8) assists in solving analytical problems, defining the problem as in section 9.2 assists in focusing your creativity and generating solutions to a design problem. One of the fears is that one cannot think of a singular idea, let alone multiple ideas for solving a problem. Ideas are very inexpensive and the greater the number of ideas generated in the beginning of a project the better, raising the probability that the final solution will be acceptable and optimal.

Roadblocks to creative problem solving spring from several sources. Past experience is valuable when it does not limit one's view in the future, when it does not become a habit. An engineer must guard against always selecting a given solution to a problem simply because it worked in the past. Preconceptions, using the same solutions, may result in failure to notice different details in a new problem. Not distinguishing between cause and effect may result in trying to solve the incorrect problem.

If your car's gas gauge is broken, the result may be that the car stops because of a lack of gasoline. However, it may also stop for other reasons—a broken distributor cap, loss of oil or loss of coolant. Only considering the lack of gas as the cause because of a preconception limits the ability to solve problems. Preconceptions block us from seeing what is.

Test yourself. Can you accurately describe the telephone in your room including the location of the numbers and letters? What about describing the dashboard of your car? Or the cover of this text?

Our personal emotions—sadness, worry and fear—are creativity roadblocks. Fear of failure, of ridicule, embarrassment, loss of a job, of appearing ignorant are tremendous negative forces. Fear often prevents people from asking questions, but asking questions is essential to the creative design process. How else can one gather data, begin to visualize and understand the problem? Ask questions and as you do so it will become easier to ask questions in the future.

When confronting a complex design problem, break it into subsystems. For instance in the design of a washing machine, there are several subassemblies that are designed separately, but co-dependently. They need to work with one another. The group or person working on the electric motor must coordinate activities with another working on the washing drum and agitator.

An interesting way to generate ideas is to use a dictionary or thesaurus. Alternate word meanings may trigger new ways of conceptualizing a solution. You may be assigned a project to create a more durable finish. Using a thesaurus, the alternate possibilities for durable include: strong, sturdy, substantial, tough, enduring, lasting, long-wearing and resistant. Each of these words brings a different nuance to mind when thinking about durable, hence this nuance may be the springboard to a new design.

It is nearly impossible to remember all thoughts and ideas in conceptualizing a new design, so keep a design notebook or journal. Record your ideas by using sketches, words, or photocopies from books and journals. It is very important to maintain a record of your activity. Sometimes personnel are reassigned in the middle of a project because the company receives another project or another engineer is assigned to work with you. The notebook becomes a way to establish project continuity. Should patents result from your work, this documentation is crucial.

During the problem definition phase, questions were asked that did not have to be answered at that time. Now you may be able to answer some of those questions. Not all questions need answering as some may require laboratory testing to resolve and in fact new ones may arise throughout the design process. Asking questions can be a stimulus to creating. For instance, what materials can be used (polymers, wood, ceramics); can a feature be enlarged (greater capacity, increased weight); can the product be made more compact, longer, shorter, heavier, lighter?

It should be noted that while the design process steps are shown in a linear fashion, they are integrative; some answers occur during the problem definition phase, questions arise while solutions are being created. Maintaining a journal is the only practical way of keeping the project history.

You do not have to generate ideas in isolation. Brainstorming is where a group of 4–12 people of different backgrounds, but equivalent levels, share ideas and generate possible solutions. Often when people hear one idea, it generates another, an offshoot from their perspective, which in turn creates another by someone else. The group needs to be prepared in advance, so they have some background information regarding the product or project. A group leader is needed to keep a record. As the group gets started, it is vital not to judge ideas, as this blocks creative thought. Evaluation occurs later. In this regard, the participants' responses need to be short, and should not contain

A laser measurement system is necessary when calibrating circuit-board drilling machines. (Courtest of Hewlett-Packard)

too much rhetoric. After the session is over, send all participants a list of the ideas created.

Referring to the problem of moving exercise bicycle parts, possible solutions could include—hand carry, a wheelbarrow, a four-wheel dolly, a two-wheel dolly (a hand truck), a conveyor belt, an overhead conveyor, a catapult. It is important not to judge ideas as they are being created; if that had occurred some of the above solutions would not exist. Aside from the advantage of having a variety of solutions, some of the ideas from an unlikely solution may improve the final design choice.

9.4 REFINEMENT AND ANALYSIS

The analysis of solutions is another area served by your engineering education; you become proficient at mathematically modeling the various physical situations. The analysis of the model will include use of the constraints that have been developed in the problem definition. The forces acting on a structure, the electrical signal that must be interpreted, are examples of these types of constraint.

For large projects, or problems requiring advances in technology, the engineer may request information from R & D. Examples include definition of new material properties of graphite epoxy composites, or new computer designs that provide the necessary computational speed. This is the quantification of the creative solutions, the modeling, the numerical analysis portion of the design process.

When the various proposed solutions have numbers associated with them, they have been quantified to some degree. A judgment must be made as to which solution is the optimum one. Economic analysis is included; manufacturing costs and marketing strategies are all considered. We can determine the power requirements of various compressor designs, for instance, but evaluating design involves more than selecting the most efficient compressor. Our evaluation must include how the compressor fits into the company's product line, whether it competes with others or achieves aesthetic value, and other items. From such an evaluation list a solution is selected.

The created solutions require refinement and analysis with the objective of finding the one(s) that best satisfy the problem statement. There are several types of analysis that need to be performed and that the final product must satisfy.

The first and primary is function analysis. Does the idea satisfy the problem statement in terms of function—can it work? Ideas that do not meet the functional requirements can be combined with other concepts or modified. Of course, this requires time and budget concerns need to be considered as to how many ideas can be functionally evaluated. Functional evaluation includes strength of materials analysis—will the product be able to withstand the static and dynamic loads applied to it? Sometimes a basic requirement needs to be reconsidered. Perhaps an original goal is too demanding, perhaps the desired product weight is too low. Analysis may reveal that by using lightweight materials, (plastics, aluminum), the requirement cannot be met. If the standard cannot be modified, then the idea must be eliminated from further consideration.

The analysis portion of the design process will feel most familiar to students as many engineering courses focus on analysis of circuits, structures, energy systems. During this analysis phase, the engineer also considers the reliability of the design, will the product last an acceptable amount of time? Determining failure rates for the product components occurs during this phase, often using information gained from statistical data.

Once the concept has satisfied the primary analysis, that it will perform correctly, it needs to be subjected to a safety analysis. Safety issues include protection from electrical shock, sharp edges, moving parts, explosion and fire, burns from hot surfaces, caustic fluids. A gasoline powered lawn mower has a power cut-off attached to the handle so that when the handle is released the engine turns off. This attempts to prevent people from putting their hands and feet under the lawn mower when the blade is turning as it is physically impossible to hold the handle and reach under the running mower.

Making a product safe is not the only consideration. It must be comfortable to use and that is where ergonomics enters the design process. How the product affects people who use it is very important and there are issues beyond comfort to consider. Product noise level, vibration, and surface temperature, readable gauges, shape and conformance to the human body are all aspects to bear in mind.

A well-designed product will be easy to service, allowing performance of the routine maintenance that needs to be accomplished for the expected operation of the device. Many motors require annual lubrication. If the access is difficult, the annual servicing may not occur. Examine an automobile's engine. Can the routine maintenance (spark plug replacement, oil and filter change) be performed readily?

With the preliminary analysis completed on the various designs, it is possible to evaluate ideas and discard the ones that are unworkable. Usually the more radical ideas, often terrific as a springboard for refinements, are dropped because the manufacturing methodology does not exist to consider them further. Ideas that may infringe on patents, fill only a small market or require too much development time are eliminated. Some others may meet a similar fate if they: (a) are too similar in design to a competitor's; (b) require too many parts, increasing manufacturing cost and decreasing reliability; (c) are complex to manufacture; or (d) will take too long to reach the market.

The selection process is simplified by drawing up a list of advantages and disadvantages for various ideas. If we consider the bicycle part conveyor, Table 9.1 could be created.

The best choices emerge for further refinement and analysis which may include building models of the product. This is often useful in complex products where clearances, lengths and locations of wire assemblies must be considered. It may be that the best design for each component or sub-system

Table 9.1

Concept	Advantage	Disadvantage
Hand carry	Immediately available, each operator carries own, little floor space required	Operator fatigue, may drop parts, unskilled task for skilled person
Wheelbarrow	Immediately available, each operator pushes own, little floor space required	Some operator fatigue, may tip over, need supply of wheelbarrows, unskilled task
Two-wheel dolly	Immediately available, each operator pushes own, little floor space required	Parts may slip off, unskilled task, supply of dollys needed
Overhead conveyor	Continuous operation, load from many locations, operator not needed	Parts may fall off, dangerous to pedestrians, parts may bump each other
Four-wheel dolly	Immediately available, each operator pushes own, more floor space required	Parts may bump, supply needed, unskilled task
Catapult	Very fast	Part damage, flying parts are dangerous, large clear space
Conveyor belt	Continuous operation, load from many locations, operator not needed	Parts may fall off, dangerous to pedestrians, parts may bump each other

is not the optimal overall design for the total assembly. A model also provides engineers with further information regarding the ease, or lack thereof, of the product for assembly—does the design require difficult steps in the manufacturing stage? Redesign at this stage is comparatively inexpensive.

9.5 PROBLEM RESOLUTION AND IMPLEMENTATION

Engineers make decisions throughout the course of the product design regarding which ideas to develop further, which to discard. Once the list has been narrowed to one or more acceptable product designs, a final decision needs to be made whether or not to proceed. Perhaps the project should be cancelled if none of the designs fully satisfy the original problem statement. Or perhaps the most acceptable design needs to be chosen and steps taken to manufacture that product.

It may appear strange that a project is cancelled after a good deal of creative effort has been put into it, but there are many possible reasons for this. The project's budget may have been exhausted. Without additional funding engineers are likely to be transferred to another project. This under-

The translation of a design on paper to the actual device can be surprising. This coal pulverizer is several stories high, and effects that can be ignored in small systems become important in large ones like this. (Courtesy of Public Service Electric and Gas Company)

scores the importance of keeping an accurate design notebook. Perhaps the time limit to develop the product was exceeded and the timing of the product's release is of major importance. Perhaps the technology being used by the company's engineers has been overtaken by newer technology not in use, rendering the product design non-competitive. Technology is constantly changing and engineers need to update their knowledge to remain competitive. This harks back to the requirement that throughout one's professional life there is the need for continued education.

On a more optimistic note, let us assume that more than one of the designs satisfies the problem statement and the task before us is to find the best solution. The multi-criteria decisionmaking techniques introduced in Chapter 8 can play an important role in this situation. List all the comparative criteria for the designs including such items as: size and weight specifications, manufacturing cost, ease of assembly, return on investment, shelf life, environmental soundness. To avoid redundancy, care must be exercised in not selecting criteria such as ease of assembly and the number of parts involved.

List the criteria in priority ranking and rate them between 1 and 10. Then compare and grade the various product designs between 1 and 10.

This helps focus on the best design(s). As noted earlier, the results of multi-criteria decisionmaking act as a guide, not a directive. The ranking and grading of the criteria takes time as it is often done by a team who must come to a consensus regarding the values.

Let us assume that a consensus has been reached as well, identifying the optimum design. Plans are drawn. Detail and assembly drawings are needed for the product. The material costs flow directly from these drawings as the parts are specified. At this point a check is made to see that the parts are current and not outdated. Product testing, via models, continues throughout the implementation phase, in order to revise the design should a sub-assembly not meet performance standards. Other information gained about the product's operation through testing can improve the design.

From a marketing view, the scheduling of the product's release can be very important. The detailed production activities in manufacturing flow

Automated production lines need to be periodically tested. (Courtest of Hewlett-Packard)

from information gained from the final drawings further indicating the importance of the drawings. This includes determining which machines need to be used, new ones to be purchased, and integration of this activity with the other ongoing manufacturing activities. Personnel require training in the assembly and manufacture of the product. The time and length of the workshops depend on the product's complexity. Additionally, the time for personnel training must be factored into the current manufacturing schedule and the implementation of the new schedule.

Customers require operating manuals and documentation of the product. In addition, service manuals are needed to repair and maintain the product along with a final parts listing. Marketing advertising and information relating to the product release must be created and coordinated with manufacturing.

Engineers play a pivotal role in successful product design and manufacture. Not only are they the creators of the design and analyzers of its performance, they are vital to successful manufacture and sales of the product. The process of bringing a product to manufacture with all the people involved requires communication between the groups throughout.

9.6 ADDITIONAL CONSIDERATIONS IN THE DESIGN PROCESS

Several other situations affect the design of a product or process. We bring to the problem we are trying to solve our own value systems. These value systems shape how we see the world and interpret what is important on a subjective level. Engineers must be aware of not only their cultural, regional, and social beliefs, but also those of the customer. This sensitivity is one of the reasons for taking courses in social sciences and humanities, to increase your breadth and depth of understanding of other people.

In line with this a variety of human factors should be considered in the design of a product. These ergonomic considerations deal with the human/machine interaction. Such factors as the relation of the equipment to a person's size, the ease of moving a dial or adjusting a CRT on a computer, the selection of colors that are harmonious or jarring, depending on the device's function, fit within this category. Often the arrangement of equipment, so it minimizes crowding of people and assures a safe working environment, physiologically and psychologically, is a concern of design teams.

Most often engineers work as part of a team. Usually the product that is being designed is too large for one person to design completely, so a team must work together, each having individual and collective responsibilities. Consider the design of a home refrigerator. Various components and aspects must be considered once the size is determined. These include the compressor specification, including sealing of the unit, piston size, lubrication and bearings, and refrigerant selection; the structure of the motor that drives the compressor, including the stator, rotor, poles, lubrication, and bearings; the size of piping to the condenser and to the evaporator; the size of the evaporator and condenser; the control system that actuates the motor/compressor; autodefrost capability; light and aesthetic considerations. This list could be expanded further. The point is that various individuals work on each of the components or subsystems, and then share information as they design the entire refrigerator system. Very often one engineer depends on the output from another subsystem, designed by another engineer, before he or she can complete the design of a component.

In other types of design projects, such as in large devices like ships, the components are already designed and available from manufacturers. The manufacturers provide catalogs that contain the specifications and operating characteristics of heat exchangers, motors, engines, radar, and generators. The design team in this case takes these known components and creates new systems. The engineer must be sure that the characteristics of the subsystems are compatible with one another. For instance, the operating speed of a motor should not be at the critical speed of the driven pump's rotor, or it would vibrate severely. It should be apparent that communication among design team members is very important to design and development of successful products.

The design process should not end when the product is produced. There will be redesign, improvements made in light of consumer comments, new technologies that can be incorporated. The United States has taken the lead in creating inventions, winning Nobel prizes. It has lost when it comes to innovation, those improvements to initial products that make them less costly, better quality, a better value. Table 9.2 illustrates a list of devices that were invented in the United States and the market share they lost. To regain

Table 9.2 INVENTION/INNOVATION
The Market Share History of Several Products

U.S.-invented Technology	1987 Market in Millions	U.S. Producers (%)			
		1970	1975	1980	1987
Phonographs	$ 630	90	40	30	1
Color TV	14,050	90	80	60	10
Audiotape recorder	500	40	10	10	0
Videotape recorder	2,895	10	10	1	1
Machine tool centers	485	99	97	79	35
Telephones	2,000	99	95	88	25
Semiconductors	19,100	89	71	65	64
Computers	53,500	NA	97	96	74

Source: U.S. Commerce Department

the lead in technological prowess, not only must engineers in the United States invent new products, but also the companies and the engineers must keep improving these products. Innovation is mandatory in maintaining a competitive edge with foreign manufacturers.

Another feature to engineering design is life-cycle design or designing with environmental considerations concerning the fabrication, operation and disposal of a piece of equipment or a process. The life-cycle analysis considers the product's environmental impact throughout its life (fabrication, operation and disposal) and tries to minimize it. Engineers will be answering questions such as what kind of pollution will be generated during raw materials production, materials conversion, product manufacture, product usage and finally product disposal. For instance, studies have shown that on average in the United States each person creates an average of 4.5 pounds of waste per day. Environmental product and process design should dramatically reduce this number and the negative environmental consequences this waste itself generates. The engineer will consider the use of recyclable materials early in the design process, designing the product for disassembly and recycling and finding ways to use less harmful processing chemicals.

Practicing engineers offer the following advice and insight:

"Keep in mind that the engineering profession is a business, and that your company must be profitable in order to remain viable. Make technical decisions based on good engineering economics. Never design something without evaluating alternatives or the cost. Call distributors and get pricing information before starting your design."

"Don't allow the stress and pressure of completing a job in an 'unreasonable' time cause you to perform inferior work. Take pride in whatever you are doing, research it completely and do the best you can. If you can say ' I could have done a better job', then you should have."

"While design, ethics, quality and many other technical issues have not changed in the 20 years I have been working, the topic at the forefront now is *teamwork*. Everything is complicated and the technical end is moving forword so fast that more people in your organization just need to work together. In most companies providing engineering services to clients, the services include multi-task types of design involving many personalities and disciplines."

"I am a design engineer designing in-house production machinery. It is *quite* a job. Everything is first done on computers, then metal is cut. This job is everything I thought engineering would be while in school and then some! In completing a design, I obtain quotes from outside vendors, design the full machine on a computer, obtain detailed prints from the assembly drawings and send them out for fabrication. The position has a pleasant mix of 'seat-of-your-pants' engineering and more textbook and math-based applications. The complete joy of designing and the agony of sometimes having it not work are both present."

REFERENCES

1. Walton, Joseph. *Engineering Design: From Art to Practice.* West Publishing, St. Paul, MN, 1991.
2. *Machine Design.* A weekly journal that contains a multitude of electrical, electronic and mechanical devices. Penton Publishing, Cleveland, OH.
3. Van Amerangen, C. *The Way Things Work.* Simon and Schustor, New York, 1971.

PROBLEMS

9.1. Investigate the following products, noting any design deficiencies or features that you believe should be added: vacuum cleaner, clothes drier, snow blower, popcorn popper, toaster, coffee maker, telephone answering machine.

9.2. Consider a procedure such as changing a car's tire or replacing a bicycle tire and list the necessary steps and tools needed for each step. Then devise alternate sequences for performing the same task.

9.3. Expand the following problems with more complete definitions:
   a) design a lamp;
   b) design a car jack;
   c) design a hair dryer;
   d) design an apartment building.

9.4. Develop additional definition to the following design problems:
   a) a reserved seat system for commuter trains;
   b) a manufacturing assembly line for making picture frames;
   c) a sorting machine for plastic bricks of various colors.

9.5. Divide a bicycle into as many subsystems as possible.

9.6. Divide an automatic toll booth into as many subsystems as possible.

9.7. Divide your home into as many subsystems as possible.

9.8. Provide a minimum of three possible solutions to explain the following situations:
   a) nervousness regarding a test;
   b) a car failing to start;
   c) getting to work on time.

9.9. Create additional uses for a football stadium that is used only 25 times a year.

9.10. Textbooks periodically are revised and new editions published. List ideas for using the old editions of the book.

9.11. Develop a list of ideas for solving the following problems:
   a) finding a dropped contact lens;
   b) locking all of a house's exterior doors from one location;
   c) a wake-up alarm for sleepy drivers;
   d) collecting golf balls on a driving range;
   e) filling a beverage bottle from a second-story window;
   f) turning pages of a book without using hands.

9.12. Problems 8, 9, 10, and 11 lend themselves to brainstorming. Get a group of students together and brainstorm, assigning one of the group to keep notes. Set a time limit for the session.

9.13. The following products have inherent dangers while being used. Identify the danger and develop ideas to overcome or limit it:
a) electric toaster;
b) electric hedge trimmer;
c) incandescent light bulb;
d) chain saw;
e) automatic garage door opener.

9.14. Develop a list of criteria that you would use in comparing different brands of the following products:
a) bicycle;
b) back pack;
c) portable computer;
d) tennis racket;
e) skis.

9.15. Make a list of advantages and disadvantages between a mechanical pencil and a wooden pencil.

9.16. Construct a list of advantages and disadvantages between three modes of transportation: automobile, train and airplane. Some of the factors to consider are cost, accessibility, safety, luggage storage and sleeping. The advantages and disadvantages may be altered when you consider in addition trips of 10 miles, 500 miles, 3000 miles.

9.17.

MEMORANDUM

Village of Kensington

TO: Jeremy Higgins, Staff Engineer
FROM: Joan Smith, Chief Engineer
SUBJ: Manhole Cover Removal
DATE: 16 November 1994

The road maintenance crew needs to inspect the storm and sewer drains in the Village. This requires that many manhole covers be removed. One of the workers has sprained his back removing a cover and will be out on disability leave for a month. Would you come up with a design so the road crew can lift and slide the manhole covers onto the street without exerting more than 40 pounds force? I need the design as soon as possible, but no later than next Friday.

9.18.

> **MEMORANDUM**
>
> ABC Engineering and Design
>
> TO: Marcy Davis, Project Engineer
> FROM: Jim Johnson, Chief Engineer
> SUBJ: Rural Mailbox
> DATE: 15 October 1994
>
> We just received an RFP (request for a proposal) from Albert Manufacturing concerning the redesign of rural mailboxes. They believe there is good market potential if they can come up with a better mailbox design, though the cost of the materials must still be less than $5. I want you to develop the RFP, including several design ideas. These will be preliminary designs which we can fully develop if we get the contract. The RFP is due November 15th, so I would like a rough draft by November 1st.

9.19.

> **MEMORANDUM**
>
> ABC Engineering and Design
>
> TO: Dave Roberts, Project Engineer
> FROM: Jim Johnson, Chief Engineer
> SUBJ: Paint Can Pourer
> DATE: 12 December 1994
>
> Rainbow Paint Company has contracted with us to produce a design for an attachment to paint cans so cans may be poured without paint running down their sides as so often occurs. They foresee a market primarily for one-gallon cans, the size most homeowners purchase. The final design needs to be completed by the end of February so I will need a memo concerning the preliminary designs with sketches by January 15th.

9.20.

> **MEMORANDUM**
>
> ABC Engineering and Design
>
> TO: Raphael Diaz, Project Engineer
> FROM: Jim Johnson, Chief Engineer
> SUBJ: Log Splitter
> DATE: 19 September 1994
>
> We have received a contract from Country Blacksmiths to design a log splitter that they can manufacture. The log splitter should be capable of being used by a homeowner for splitting firewood for use in fireplaces and wood stoves. They would like a simple and effective design made from steel. The company is most familiar with using this material. They requested a meeting with us the beginning of October at which time we need to present some conceptual designs. Send me a memo with some preliminary sketches by the 28th of September.

9.21. The following is a list of possible design projects:

Portable garage that can be collapsed and stored in a car's trunk.
A tray for eating and writing for rear seat passengers in an automobile.
Automobile jacks that are built into the car's body.
Heated steering wheel.
A bicycle trailer.
A car wash for pick-up trucks.
A means to alert drowsy drivers.
A mass transit system for a college campus.
A boat with retractable trailer wheels and a trailer hitch.
Improved bicycle brakes.
A soap dispenser.
A shoestring fastener which replaces tying laces.
An automatic car door opener.
A soda/beer can crusher.
An impact hammer for use on electric drills.
A carbon monoxide indicator.
A bookshelf that adjusts vertically and horizontally.
A compact and portable stove for backpacking.

Appendices

Appendix 1   NSPE Code of Ethics for Engineers

Appendix 2   Conversion Factors

Appendix 3   Computers and Computer Applications

Appendix 4   Algebraic and Trigonometric Problem Solving

Appendix One

National Society of Professional Engineers Code of Ethics for Engineers

PREAMBLE

Engineering is an important and learned profession. The members of the profession recognize that their work has a direct and vital impact on the quality of life for all people. Accordingly, the services provided by engineers require honesty, impartiality, fairness and equity, and must be dedicated to the protection of the public health, safety and welfare. In the practice of their profession, engineers must perform under a standard of professional behavior which requires adherence to the highest principles of ethical conduct on behalf of the public, clients, employers and the profession.

FUNDAMENTAL CANONS

Engineers, in the fulfillment of their professional duties, shall:

1. Hold paramount the safety, health and welfare of the public in the performance of their professional duties.
2. Perform services only in areas of their competence.
3. Issue public statements only in an objective and truthful manner.
4. Act in professional matters for each employer or client as faithful agents or trustees.
5. Avoid improper solicitation of professional employment.

RULES OF PRACTICE

1. Engineers shall hold paramount the safety, health and welfare of the public in the performance of their professional duties.
   (a) Engineers shall at all times recognize that their primary obligation is to protect the safety, health, property and welfare of the public. If their professional judgment is overruled under circumstances where the safety, health, property or welfare of the public are endangered, they shall notify their employer or client and such other authority as may be appropriate.
   (b) Engineers shall approve only those engineering documents which are safe for public health, property and welfare in conformity with accepted standards.
   (c) Engineers shall not reveal facts, data or information obtained in a professional capacity without the prior consent of the client or employer except as authorized or required by law or this Code.
   (d) Engineers shall not permit the use of their name or firm name nor associate in business ventures with any person or firm which they have reason to believe is engaging in fraudulent or dishonest business or professional practices.
   (e) Engineers having knowledge of alleged violations of this Code shall cooperate with the proper authorities in furnishing such information or assistance as may be required.
2. Engineers shall perform services only in the areas of their competence.
   (a) Engineers shall undertake assignments only when qualified by education or experience in the specific technical field involved.
   (b) Engineers shall not affix their signatures to any plans or documents dealing with subject matter in which they lack competence, nor to any plan or document not prepared under their direction and control.
   (c) Engineers may accept an assignment outside of their fields of competence to the extent that their services are restricted to those phases of the project in which they are qualified, and to the extent that they are satisfied that all other phases of such project will be performed by registered or otherwise qualified associates, consultants, or employees, in which case they may then sign the documents for the total project.
3. Engineers shall issue public statements only in an objective and truthful manner.
   (a) Engineers shall be objective and truthful in professional reports, statements or testimony. They shall include all rele-

vant and pertinent information in such reports, statements or testimony.

(b) Engineers may express publicly a professional opinion on technical subjects only when the opinion is founded upon adequate knowledge of the facts and competence in the subject matter.

(c) Engineers shall issue no statements, criticims or arguments on technical matters which are inspired or paid for by interested parties, unless they have prefaced their comments by explicitly identifying the interested parties on whose behalf they are speaking, and by revealing the existence of any interest the engineers may have in the matters.

4. Engineers shall act in professional matters for each employer or client as faithful agents or trustees.

(a) Engineers shall disclose all known or potential conflicts of interest to their employers or clients by promptly informing them of any business association, interest or other circumstances which could influence or appear to influence their judgment or the quality of their services.

(b) Engineers shall not accept compensation, financial or otherwise, from more than one party for services on the same project, or for services pertaining to the same project, unless the circumstances are fully disclosed to, and agreed to, by all interested parties.

(c) Engineers shall not solicit or accept finanical or other valuable consideration, directly or indirectly, from contractors, their agents, or other parties in connection with work for employers or clients for which they are responsible.

(d) Engineers in public service as members, advisors or employees of a governmental body or department shall not participate in decisions with respect to professional services solicited or provided by them or their organizations in private or public engineering practice.

(e) Engineers shall not solicit or accept a professional contract from a governmental body on which a principal or officer of their organization serves as a member.

5. Engineers shall avoid improper solicitation of professional employment.

(a) Engineers shall not falsify or permit misrepresentation of their, or their associates', academic or professional qualifications. They shall not misrepresent or exaggerate their degree of responsibility in or for the subject matter of prior assignments. Brochures or other presentations incident to the solicitation of employment shall not misrepresent pertinent facts concerning employers, employees, associates, joint ventures or past accomplishments with the intent and purpose of enhancing their qualifications and their work.

(b) Engineers shall not offer, give, solicit or receive, either directly or indirectly, any political contribution in an amount intended to influence the award of a contract by public authority, or which may be reasonably construed by the public of having the effect or intent to influence the award of a contract. They shall not offer any gift, or other valuable consideration, in order to secure work. They shall not pay a commission, percentage or brokerage fee in order to secure work except to a bona fide employee or bona fide established commercial or marketing agencies retained by them.

PROFESSIONAL OBLIGATIONS

1. Engineers shall be guided in all their professional relations by the highest standards of integrity.
   - **(a)** Engineers shall admit and accept their own errors when proven wrong and refrain from distorting or altering the facts in an attempt to justify their decisions.
   - **(b)** Engineers shall advise their clients or employers when they believe a project will not be successful.
   - **(c)** Engineers shall not accept outside employment to the detriment of their regular work or interest. Before accepting any outside employment they will notify their employers.
   - **(d)** Engineers shall not attempt to attract an engineer from another employer by false or misleading pretenses.
   - **(e)** Engineers shall not actively participate in strikes, picket lines, or other collective coercive action.
   - **(f)** Engineers shall avoid any act tending to promote their own interest at the expense of the dignity and integrity of the profession.
2. Engineers shall at all times strive to serve the public interest.
   - **(a)** Engineers shall seek opportunities to be of constructive service in civic affairs and work for the advancement of the safety, health and well-being of their community.
   - **(b)** Engineers shall not complete, sign or seal plans and/or specifications that are not of a design safe to the public health and welfare and in conformity with accepted engineering standards. If the client or employer insists on such unprofessional conduct, they shall notify the proper authorities and withdraw from further services on the project.
   - **(c)** Engineers shall endeavor to extend public knowledge and appreciation of engineering and its achievements and to protect the engineering profession from misrepresentation and misunderstanding.
3. Engineers shall avoid all conduct or practice which is likely to discredit the profession or deceive the public.

(a) Engineers shall avoid the use of statements containing a material misrepresentation of fact or omitting a material fact necessary to keep statements from being misleading; statements intended or likely to create an unjustified expectation; statements containing prediction of future success; statements containing an opinion as to the quality of the engineers' services; or statements intended or likely to attract clients by the use of showmanship, puffery, or self-laudation, including the use of slogans, jingles, or sensational language or format.
(b) Consistent with the foregoing; engineers may advertise for recruitment of personnel.
(c) Consistent with the foregoing; engineers may prepare articles for the lay or technical press, but such articles shall not imply credit to the author for work performed by others.

4. Engineers shall not disclose confidential information concerning the business affairs or technical processes of any present or former client or employer without his consent.
    (a) Engineers in the employ of others shall not without the consent of all interested parties enter promotional efforts or negotiations for work or make arrangements for other employment as a principal or to practice in connection with a specific project for which the engineer has gained particular and specialized knowledge.
    (b) Engineers shall not, without the consent of all interested parties, participate in or represent an adversary interest in connection with a specific project or proceeding in which the engineer has gained particular specialized knowledge on behalf of a former client or employer.

5. Engineers shall not be influenced in their professional duties by conflicting interests.
    (a) Engineers shall not accept financial or other considerations, including free engineering designs, from material or equipment suppliers for specifying their product.
    (b) Engineers shall not accept commissions or allowances, directly or indirectly, from contractors or other parties dealing with clients or employers of the engineer in connection with work for which the engineer is responsible.

6. Engineers shall uphold the principle of appropriate and adequate compensation for those engaged in engineering work.
    (a) Engineers shall not accept remuneration from either an employee or employment agency for giving employment.
    (b) Engineers, when employing other engineers, shall offer a salary according to professional qualifications and the recognized standards in the particular geographical area.

7. Engineers shall not compete unfairly with other engineers by attempting to obtain employment or advancement or professional engagements by taking advantage of a salaried position, by criticizing other engineers, or by other improper or questionable methods.
   (a) Engineers shall not request, propose, or accept a professional commission on a contingent basis under circumstances in which their professional judgment may be compromised.
   (b) Engineers in salaried positions shall accept part-time engineering work only at salaries not less than that recognized as standard in the area.
   (c) Engineers shall not use equipment, supplies, laboratory or office facilities of an employer to carry on outside private practice without consent.
8. Engineers shall not attempt to injure, maliciously or falsely, directly or indirectly, the professional reputation, prospects, practice or employment of other engineers, nor indiscriminately criticize other engineers' work. Engineers who believe others are guilty of unethical or illegal practice shall present such information to the proper authority for action.
   (a) Engineers in private practice shall not review the work of another engineer for the same client, except with the knowledge of such engineer, or unless the connection of such engineer with the work has been terminated.
   (b) Engineers in governmental, industrial or educational employ are entitled to review and evaluate the work of other engineers when so required by their employment duties.
   (c) Engineers in sales or industrial employ are entitled to make engineering comparisons of represented products with products of other suppliers.
9. Engineers shall accept personal resposibility for all professional activities.
   (a) Engineers shall conform with state registration laws in the practice of engineering.
   (b) Engineers shall not use association with a nonengineer, a corporation or a partnership, as a "cloak" for unethical acts, but must accept personal responsibility for all professional acts.
10. Engineers shall give credit for engineering work to those to whom credit is due, and will recognize the proprietary interests of others.
    (a) Engineers shall, whenever possible, name the person or persons who may be individually responsible for designs, inventions, writings or other accomplishments.

- (b) Engineers using designs supplied by a client recognize that the designs remain the property of the client and may not be duplicated by the engineer for others without express permission.
- (c) Engineers, before undertaking work for others in connection with which the engineer may make improvements, plans, designs, inventions, or other records which may justify copyright or patents, should enter into a positive agreement regarding ownership.
- (d) Engineers' designs, data, records, and notes referring exclusively to an employer's work are the employer's property.

11. Engineers shall cooperate in extending the effectiveness of the profession by interchanging information and experience with other engineers and students, and will endeavor to provide opportunity for professional development and advancement of engineers under their supervision.
    - (a) Engineers shall encourage engineering employee's efforts to improve their education.
    - (b) Engineers shall encourage engineering employees to attend and present papers at professional and technical society meetings.
    - (c) Engineers shall urge engineering employees to become registered at the earliest possible date.
    - (d) Engineers shall assign a professional engineer duties of a nature to utilize full training and experience, insofar as possible, and delegate lesser functions to subprofessionals or to technicians.
    - (e) Engineers shall provide a prospective engineering employee with complete information on working conditions and proposed status of employment, and after employment will keep employees informed of any changes.

AS REVISED MARCH 1985

"By order of the United States District Court for the District of Columbia, former Section 11(c) of the NSPE Code of Ethics prohibiting competitive bidding, and all policy statements, opinions, rulings or other guidelines interpreting its scope, have been rescinded as unlawfully interfering with the legal right of engineers, protected under the antitrust laws, to provide price information to prospective clients; accordingly, nothing contained in the NSPE Code of Ethics, policy statements, opinions, rulings or other guidelines, prohibits the submission of price quotations or competitive bids for engineering services at any time or in any amount."

STATEMENT BY NSPE EXECUTIVE COMMITTEE

In order to correct misunderstandings which have been indicated in some instances since the issuance of the Supreme Court decision and the entry of the final judgment, it is noted that in its decision of April 25, 1978, the Supreme Court of the United States declared: "The Sherman Act does not require competitive bidding."

It is further noted that as made clear in the Supreme Court decision:

1. Engineers and firms may individually refuse to bid for engineering services.
2. Clients are not required to seek bids for engineering services.
3. Federal, state and local laws governing procedures to procure engineering services are not affected, and remain in full force and effect.
4. State societies and local chapters are free to actively and aggressively seek legislation for professional selection and negotation procedures by public agencies.
5. State registration board rules of professional conduct, including rules prohibiting competitive bidding for engineering services, are not affected and remain in full force and effect. State registration boards with authority to adopt rules of professional conduct may adopt rules governing procedures to obtain engineering services.
6. As noted by the Supreme Court, "nothing in the judgment prevents NSPE and its members from attempting to influence governmental action. . . ."[1]

Note: In regard to the question of application of the Code to corporations vis-à-vis real persons, business form or type should not negate nor influence conformance of individuals to the Code. The Code deals with professional services, which services must be performed by real persons. Real persons in turn establish and implement policies within business structures. The Code is clearly written to apply to the engineer and it is incumbent on a member of NSPE to endeavor to live up to its provisions. This applies to all pertinent sections of the Code.

---

[1]*Source*: NPSE Publication No. 1102 as revised January 1987

Appendix Two

Conversion Factors

Multiply	By	To Obtain
abamperes	10	amperes
abcoulombs	10	coulombs
abfarads	10^9	farads
abhenries	10^{-9}	henries
abohms	10^{-9}	ohms
abvolts	10^{-8}	volts
acres	43 560	square feet
"	4047	square meters
"	4840	square yards
acre-feet	43 560	cubic feet
" "	3.259×10^5	gallons (U.S. liquid)
amperes	1	coulombs per second
angstroms	10^{-10}	meters
ares	0.02471	acres
"	100	square meters
astronomical units	1.496×10^{11}	meters
atmospheres	76	centimeters of mercury
"	29.92	inches of mercury
"	33.90	feet of water
"	14.70	pounds per square inch (force)
"	1.013×10^5	pascals
bars	0.9872	atmospheres
"	10^6	dynes per square centimeter
"	14.51	pounds per square inch (force)
"	10^5	pascals
barrels (petroleum)	42	gallons (U.S. liquid)
becquerels (radioactivity)	1	disintegrations per second

Multiply	By	To Obtain
British thermal units	778.2	foot-pounds force
" " "	1055	joules
" " "	2.931×10^{-4}	kilowatt hours
Btu per minute	17.58	watts
Btu per pound	2326	joules per kilogram
bushels	1.244	cubic feet
"	0.035 24	cubic meters
"	4	pecks
calories	4.187	joules
"	10^{-3}	kilocalories
calories per gram	4187	joules per kilogram
Calories (kilocalories)	4187	joules
candelas per square meter	3.141×10^{-4}	lamberts
carats (metric)	2×10^{-4}	kilograms (mass)
centimeters	0.3937	inches
centimeters of mercury	0.013 16	atmospheres
" " "	0.4461	feet of water
" " "	0.1935	pounds per square inch (force)
" " "	1333	pascals
circular mils	5.067×10^{-6}	square centimeters
" "	7.854×10^{-7}	square inches
cords	8 ft. $\times$ 4 ft. $\times$ 4 ft.	cubic feet
coulombs (quantity of electricity)	1	ampere-seconds
cubic centimeters	3.531×10^{-5}	cubic feet
" "	6.102×10^{-2}	cubic inches
" "	10^{-6}	cubic meters
" "	10^{-3}	liters
" "	2.642×10^{-4}	gallons (U.S. liquid)
cubic feet	2.832×10^{4}	cubic centimeters
" "	1728	cubic inches
" "	0.028 32	cubic meters
" "	0.037 04	cubic yards
" "	7.481	gallons (U.S. liquid)
" "	28.32	liters
cubic inches	16.39	cubic centimeters
" "	5.787×10^{-4}	cubic feet
" "	1.639×10^{-5}	cubic meters
" "	2.143×10^{-5}	cubic yards
" "	4.329×10^{-3}	gallons (U.S. liquid)
cubic meters	35.31	cubic feet

CONVERSION FACTORS

Multiply	By	To Obtain
cubic meters	61 024	cubic inches
" "	1.308	cubic yards
" "	264.2	gallons (U.S. liquid)
cubic yards	27	cubic feet
" "	46 656	cubic inches
" "	0.7646	cubic meters
" "	202.0	gallons (U.S. liquid)
curies	3.7×10^{10}	becquerels
days	24	hours
"	1440	minutes (time)
"	8.640×10^4	seconds (time)
degrees (angle)	60	minutes (angle)
" "	0.017 45	radians
degrees Fahrenheit	—	degrees Celsius: $t_C = (t_F - 32)/1.8$
degrees Celsius	—	kelvin: $T = t_C + 273.15$ K
degrees Fahrenheit	—	kelvin: $T = (t_F + 459.67°R)/1.8$
degrees Rankine	—	kelvin: $T = T_R/1.8$
degrees per second (angle)	0.1667	revolutions per minute
degrees Kelvin (see kelvin)		
density: pounds-mass/in.^{3}	27 680	kilograms per cubic meter (mass)
drams	1.772	grams force
"	0.0625	ounces force
dynes	1.020×10^{-3}	grams force
"	7.233×10^{-5}	poundals
"	2.248×10^{-6}	pounds force
"	1	gram-centimeters/s^2 (mass)
"	10^{-5}	newtons
electron volts	1.602×10^{-19}	joules
ergs	9.479×10^{-11}	British thermal units
"	7.378×10^{-8}	foot-pounds force
"	10^{-7}	joules
"	1	dyne-centimeters
ergs per second	1.341×10^{-10}	horsepower
" " "	10^{-7}	watts
farads (electric capacitance)	1	coulombs per volt
fathoms	6	feet

Multiply	By	To Obtain
feet	0.3048	meters
feet per second	0.3048	meters per second
feet of water	0.029 50	atmospheres
″ ″ ″	0.8827	inches of mercury
″ ″ ″	0.4336	pounds per square inch (force)
feet of water (39.2°F)	2989	pascals
foot-candles	10.76	lumens per square meter (lux)
″ ″	10.76	lux
″ ″	1	lumens per square foot
foot-pounds force	1.285×10^{-3}	British thermal units
″ ″ ″	1.356×10^{7}	ergs
″ ″ ″	1.356	joules
force: lbf	4.448	newtons
force: kgf ("kilopond")	9.807	newtons
force (1 kg · m/s^2)	1	newtons
frequency (1/s)	1	hertz
furlongs	40	rods
gallons (U.S. liquid)	3.785×10^{-3}	cubic meters
″ ″ ″	0.1337	cubic feet
″ ″ ″	231	cubic inches
″ ″ ″	4	quarts (U.S. liquid)
gallons (U.S. dry)	4.405×10^{-3}	cubic meters
gallons (U.K. liquid)	4.546×10^{-3}	cubic meters
gals (unit of acceleration)	10^{-2}	meters per second per second
gammas (mass)	10^{-9}	kilograms (mass)
gammas (magnetic flux density)	10^{-9}	teslas
gausses	10^{-4}	teslas
gills	0.25	pints (U.S. liquid)
grads	1.571×10^{-2}	radians
grains	1.429×10^{-4}	pounds
grams	10^{-3}	kilograms
grams force	0.035 27	ounces force
″ ″	0.032 15	ounces force (troy)
″ ″	2.204×10^{-3}	pounds force
hectares	2.471	acres
″	10^{4}	square meters
henries (inductance)	1	webers per ampere
horsepower	42.41	British thermal units per minute

Multiply	By	To Obtain
horsepower	33 000	foot-pounds per minute (force)
″ ″	550	foot-pounds per second (force)
″ ″	747.7	watts
horsepower-hour	2.684×10^6	joules
inches	2.540	centimeters
″	2.540×10^{-2}	meters
inches of mercury (32°F)	0.033 42	atmospheres
″ ″ ″ ″	0.4912	pounds per square inch (force)
″ ″ ″ ″	3.386×10^3	pascals
inches of mercury (60°F)	3.377×10^3	pascals
joules (energy, work, heat)	1	newton meters
″	9.478×10^{-4}	British thermal units
″	0.7376	foot-pounds force
″	2.778×10^{-4}	watt-hours
″	0.2388	calories
joules	2.388×10^{-4}	kilocalories
joules per kilogram	4.300×10^{-4}	Btu per pound
kelvin	—	degrees Celsius: $t_C = T - 273.15$ K
″	—	degrees Fahrenheit: $t_F = 1.8\,T - 459.67$ R
″	—	degrees Rankine: $T_R = 1.8\,T_K$
kilocalories	4.187×10^3	joules
″	10^3	calories
kilograms force (kgf)	70.93	poundals
″ ″ ″	2.205	pounds force
″ ″ ″	9.807	newtons
kilograms mass (kg)	1	kilograms
″ ″ ″	0.068 54	slugs (mass)
″ ″ ″	2.205	pounds mass
kilograms per cubic meter	0.062 43	pounds per cubic foot
kilograms per square meter (force)	1.422×10^{-3}	pounds per square inch (force)
kilometers	3281	feet
″	0.6214	miles
″	10^3	meters
kiloponds (kgf)	9.807	newtons
kilowatts	10^3	watts
kilowatt-hours	3.600×10^6	joules
kips (1000 lbf)	4.448×10^3	newtons

Multiply	By	To Obtain
kips per square inch	6.895×10^6	pascals
knots (international)	1.151	miles per hour
" "	0.5144	meters per second
lamberts	3183	candelas per square meter
leagues (neutical)	5556	meters
leagues (U.S. survey)	4828	meters
light years	9.461×10^{15}	meters
liters	10^{-3}	cubic meters
"	0.035 31	cubic feet
"	0.2642	gallons (U.S. liquid)
"	10^3	cubic centimeters
lumens (luminous flux)	1	candela-steradians
lumens per square foot	1	foot-candles
lumens per square meter	1	lux
lux (illuminance)	1	lumens per square meter
lux (lm/m^2)	0.0929	foot-candles
mass: lbm	0.4536	kilograms (mass)
maxwells	10^{-8}	webers
meters	1.094	yards
"	3.281	feet
meters	39.37	inches
"	6.214×10^{-4}	miles (U.S. survey)
meters per second	3.281	feet per second
metric carats	2×10^{-4}	kilograms
metric tons (tonnes)	10^3	kilograms
mhos	1	siemens
microns	10^{-6}	meters
miles (nautical)	1852	meters
miles (U.S. survey)	1609	meters
" " "	5280	feet
" " "	1.609	kilometers
" " "	1760	yards
miles per hour	88	feet per minute
" " "	0.8688	knots (international)
milliamperes	10^{-3}	amperes
millibars	10^2	pascals
millimeters	0.039 37	inches
"	10^{-3}	meters
millimeters of mercury (0°C)	133.3	pascals

Multiply	By	To Obtain
millivolts	10^{-3}	volts
mils	10^{-3}	inches
miner's inches	1.5	cubic feet per minute
minutes (angle)	2.909×10^{-4}	radians
nautical miles	1852	meters
newtons	1	kilogram-meters per second per second (kg · m/s^2)
"	0.2248	pounds-force
"	10^5	dynes
"	0.1020	kilograms force
"	7.233	poundals
newton meters	0.7376	pounds-feet (force)
oersteds	79.58	amperes per meter
ohms (electric resistance)	1	volts per ampere
ounces (troy)	0.083 33	pounds (troy)
" "	1.097	ounces (avoirdupois)
ounces force	0.2780	newtons
" "	28.35	grams force
" "	0.0625	pounds force
ounces force (troy)	31.10	grams force
parsecs	3.084×10^{16}	meters
pascals (pressure, stress)	1	newtons per square meter
"	0.9872×10^{-5}	atmospheres
"	2.953×10^{-4}	inches of mercury (32°F)
"	7.501×10^{-3}	millimeters of mercury (0°C) (torr)
pascals	1.450×10^{-4}	pounds per square inch (force)
pecks (U.S.)	8.810×10^{-3}	cubic meters
pennyweights	$1\,555 \times 10^{-3}$	kilograms (mass)
picas (printer's)	4.218×10^{-3}	meters
pints (U.S. liquid)	4.732×10^{-4}	cubic meters
" (U.S. dry)	5.506×10^{-4}	cubic meters
points (printer's)	3.515×10^{-4}	meters
poises (absolute viscosity)	10^{-1}	pascal-seconds
poundals	0.1383	newtons
"	1.383×10^4	dynes
"	0.031 08	pounds force
pounds (avoirdupois)	7000	grains
pounds (troy)	0.8229	pounds (avoirdupois)
" "	5760	grains

Multiply	By	To Obtain
pound-feet (force)	1.356	newton meters
pounds force (lbf)	453.7	grams force
" " "	16	ounces force
" " "	32.18	poundals
" " "	4.448	newtons
pounds mass (lbm)	0.4536	kilograms (mass)
pounds per cu ft	16.02	kilograms per cubic meter
pounds per sq in (psi)	0.068 03	atmospheres
" " " "	2.036	inches of mercury
" " " "	6895	pascals
" " " "	6.895×10^{-3}	megapascals
pressure: psi	6895	pascals
pressure: atmospheres	1.013×10^5	pascals
quarts (U.S. liquid)	9.464×10^{-4}	cubic meters
" " "	0.2500	gallons (U.S. liquid)
radians	57.30	degrees (angle)
"	63.65	grads
"	0.1592	revolutions
rads (radiation dose absorbed)	10^{-2}	joules per kilogram (grays)
rods (U.S. survey)	16.5	feet
roentgens	2.580×10^{-4}	coulombs per kilogram
revolutions	2π	radians
sections (U.S. survey)	640	acres
"	2.590×10^6	square meters
siemens (electric conductance)	1	amperes per volt
"	1	mhos
slugs (mass)	14.59	kilograms (mass)
square centimeters	10^{-4}	square meters
statamperes	3.336×10^{-10}	amperes
statcoulombs	3.336×10^{-10}	coulombs
statfarads	1.113×10^{-12}	farads
stathenries	8.988×10^{11}	henries
statohms	8.988×10^{11}	ohms
statvolts	299.8	volts
steres	1	cubic meters
stokes (kinematic viscosity)	10^{-4}	square meters per second
tablespoons	1.479×10^{-5}	cubic meters
teaspoons	4.929×10^{-6}	cubic meters
temperature (°C) + 273.15	1	absolute temperature (kelvin)

CONVERSION FACTORS

Multiply	By	To Obtain
temperature (°C) + 17.78	1.8	temperature (°F)
temperature (°F) + 459.67	1	absolute temperature (°R)
" "" − 32	5/9	temperature (°C)
teslas (magnetic flux density)	1	webers per square meter
teslas	10^4	gausses
therms	10^5	British thermal units (Btu)
tonnes (metric tons)	10^3	kilograms
tons (long)	1016	kilograms
" "	2240	pounds
tons (metric)	10^3	kilograms
" "	2205	pounds
tons (short)	907.2	kilograms
" "	2000	pounds
tons (of refrigeration)	1.2×10^4	British thermal units per hour
tons (nuclear equivalent of TNT)	4.20×10^9	joules
torr (mm Hg, 0°C)	133.3	pascals
volts (electric potential)	1	watts per ampere
watts (power)	0.056 88	British thermal units per minute
" "	10^7	ergs per second
" "	1.341×10^{-3}	horsepower
" "	1	joules per second
watt-hours	3600	joules
webers (magnetic flux)	1	volt-seconds
"	10^8	maxwells
yards	0.9144	meters
"	3	feet

Appendix Three

Computers and Computer Applications

Computers characterize our current society in much the same way that factories characterized the earlier industrial age. When one thinks about the industrial era, images of factories and smokestacks emerge. The postindustrial era, often called the computer era, is characterized by computers influencing many aspects of our lives, the types of services available to us, and the technologies we use. Engineers view computers in at least two ways: how to use them as tools in solving problems in a variety of engineering disciplines, and how they function, at least conceptually.

A3.1 THE COMPUTER PROCESSING CYCLE

Computers perform simple operations with tremendous speed and accuracy. Information enters the computer through an input device, such as the keyboard at a computer terminal. The computer performs various calculations with these data in the central processing unit (CPU), and the results are passed to an output unit, such as a printer or video display. Thus, for an engineering report you input the keystrokes for the various words, the CPU translates these into binary numbers and then back into letters, which are displayed on the video terminal. For another part of the report you may input experimental data; the CPU will perform calculations using the data and send the results of the calculations to the printer.

Figure A3.1 illustrates the various components in a computer system. The input and output devices include tape drives, keyboards, disk drives, printers, card readers, card punches, video terminals, and speakers. The heart of the computer is the CPU. It consists of three units, the processing unit, more typically called the arithmetic logic unit (ALU), the main storage unit, and the control unit.

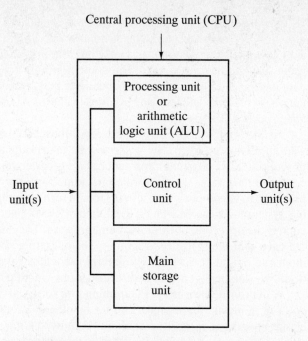

Figure A3.1 The basic components in a computer system

The ALU performs all the arithmetical calculations and makes logical decisions as well, such as whether or not a number is greater than the previous one. How does the ALU receive the information in the first place? When you input the data the control unit directs this information to the storage unit so it is available when the ALU needs it for calculations. The storage unit then holds the results of the calculations for the output unit. All these functions can take place on the surface of a microchip.

The ALU performs all the additions, subtractions, multiplications, and divisions as well as logical decision making. The computer decision-making capabilities are built-in, or rather programmed in machine language instructions, and do not require additional programming by the user. Three of the typical comparative operations are the "equal to," "greater than," and "less than" operations.

How do you determine if an item is stocked in a warehouse's inventory? Enter the item's identification code and the computer searches for a number equal to this; if found, then the item is in the inventory. Credit cards are frequently used to purchase items in a store. As part of the charging process, the store will check your card number with a central computer to determine if the available balance is greater than the amount you wish to charge. The "less than" function can be used in situations requiring that a certain minimum quantity be met, such as in a temperature-sensing system.

How does the computer know to process the data? The set of instructions directing this is in the computer program. The programs are held in main storage enabling the computer to execute the instructions when necessary, using the data held in storage. The computer must keep track of where the data are located and has memory locations that designate where a piece of data is located, its address, and what its value is. Figure A3.2 diagrams the two aspects of the memory location.

As a computer user, the engineer will most frequently be running software to solve a problem. He or she will be interfacing with the applications program, a CAD program for instance. This program must eventually use the computer hardware, the electronic bits, to perform the various operations it requires. To accomplish this, it interfaces in turn with the computer's operating system, such as MS-DOS, UNIX, or CP/M. These operating systems control the operation of the computer hardware. They manage the hardware, the data, and the software programs. The operating system also directs the input and output devices, allows several users to access the same computer, or allows one person to run several programs simultaneously.

A microprocessor is a CPU in miniature form, fabricated on an IC chip encased in plastic for strength and for holding the connecting pins. Printed circuit boards are comprised of many IC chips that may include a microprocessor. As you are undoubtedly aware, circuit boards are used in a variety of applications, from automobile control systems to children's toys.

Microcomputers commonly have one of two types of memory, ROM (read-only memory) and RAM (random access memory). Once they are fabricated, ROM memories cannot be altered, but neither can their information be lost, and it is instantly available when the power is turned on. They are not affected by loss of power as RAM memories are. They are found in controllers, such as those used in microwave ovens and in calculator functions. Some of the disadvantages of ROM are that the memory cannot be used for other purposes, it cannot be written into, and it is not easily programmed except at the factory, where the programming is performed on a large volume basis. One type of ROM can be programmed by the purchaser, but it remains permanent after it is programmed. This is called PROM (programmable read-only memory). Not to be defeated by a PROM that couldn't be repro-

Address	Value
0	−16.21
1	0.1695
2	1000.1
3	−24

Figure A3.2 Computer memory location elements.

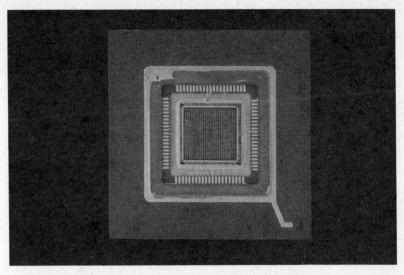

A computer chip. (Courtesy of Grumman Aerospace Corporation)

grammed, electrical engineers created an EPROM (erasable programmable read-only memory), which can be reprogrammed using special procedures, such as erasing the EPROM with ultraviolet light and then reprogramming the chip.

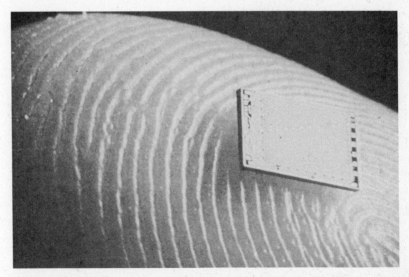

A microprocessor is contained on a single silicon chip, small enough to fit on a fingertip. (Courtesy of ASME)

A common place for ROMs is in calculators for use by special function keys, such as the exponential function. The function may be written as an infinite series, such as

$$e^x = 1 + x + \frac{x^2}{2!} + \frac{x^3}{3!} + \frac{x^4}{4!} + \cdots$$

When the key is pushed, the exponential is calculated using the ROM program and the result displayed. This explains the time delay from the start of the execution until the number is displayed on the calculator.

RAM memories are programmable, a singular advantage, but they are susceptible to power loss, a singular disadvantage. The program size that a microcomputer can execute depends on the RAM available. Microcomputers will have both RAM and ROM memories to serve different functions.

Engineering calculators have many features such as entering equations as they appear in texts and manipulating graphs. (Courtesy of Hewlett-Packard)

A3.2 COMPUTER APPLICATIONS

Perhaps the area of greatest impact that the computer has had is in the applications that we can now perform using it. Imagine a circle and smaller circles around its perimeter and tangent to it. The inner circle represents a computer, and the outer circles represent applications using the computer, such as word processing, computer-aided design, computer-aided manufacturing, spreadsheets, data-base management, desktop publishing, finite element analysis, and robotics. Each of the applications is linked to the others through the computer. As the power and utility of the computer increases, its "diameter" increases and new circles of applications fit on the perimeter.

The computer affects our lives and work in myriad ways, perhaps none more fundamentally important than the way goods are manufactured. Before the advent of the industrial age, people handcrafted devices; they literally made them. With the industrial revolution, machines made the devices; people operated the machines. Now people supervise a computer, which operates a machine, which makes the device. This occurs daily as computers operate automated bank tellers, robots, and numerically controlled machines.

The role of people in the computer age is that of supervision, contrasted with operation in the earlier industrial age. As such, engineers use a variety of software packages to supervise the computer's operation and need to know how information is transmitted between programs and within programs for proper supervision. Several codes have been established for this communication. The most popular is an alphanumeric code called ASCII (American Standard Code for Information Interchange). It is a seven-bit code that allows

The X-29 is a forward swept wing experimental aircraft. The aircraft requires computer control, as the human response time of the pilot is not quick enough to guide the plane. The forward swept wings allow better performance at supersonic speeds than the traditional rear swept wing design. (Courtesy of Grumman Aerospace Corporation)

Table A3.1 ASCII CODE CHARACTERS AND THEIR BIT EQUIVALENTS

Character	ASCII Code	Character	ASCII Code	Character	ASCII Code
space	0 1 0 0 0 0 0	@	1 0 0 0 0 0 0		1 1 0 0 0 0 0
!	0 1 0 0 0 0 1	A	1 0 0 0 0 0 1	a	1 1 0 0 0 0 1
"	0 1 0 0 0 1 0	B	1 0 0 0 0 1 0	b	1 1 0 0 0 1 0
#	0 1 0 0 0 1 1	C	1 0 0 0 0 1 1	c	1 1 0 0 0 1 1
$	0 1 0 0 1 0 0	D	1 0 0 0 1 0 0	d	1 1 0 0 1 0 0
%	0 1 0 0 1 0 1	E	1 0 0 0 1 0 1	e	1 1 0 0 1 0 1
&	0 1 0 0 1 1 0	F	1 0 0 0 1 1 0	f	1 1 0 0 1 1 0
'	0 1 0 0 1 1 1	G	1 0 0 0 1 1 1	g	1 1 0 0 1 1 1
(	0 1 0 1 0 0 0	H	1 0 0 1 0 0 0	h	1 1 0 1 0 0 0
)	0 1 0 1 0 0 1	I	1 0 0 1 0 0 1	i	1 1 0 1 0 0 1
*	0 1 0 1 0 1 0	J	1 0 0 1 0 1 0	j	1 1 0 1 0 1 0
+	0 1 0 1 0 1 1	K	1 0 0 1 0 1 1	k	1 1 0 1 0 1 1
,	0 1 0 1 1 0 0	L	1 0 0 1 1 0 0	l	1 1 0 1 1 0 0
−	0 1 0 1 1 0 1	M	1 0 0 1 1 0 1	m	1 1 0 1 1 0 1
.	0 1 0 1 1 1 0	N	1 0 0 1 1 1 0	n	1 1 0 1 1 1 0
/	0 1 0 1 1 1 1	O	1 0 0 1 1 1 1	o	1 1 0 1 1 1 1
0	0 1 1 0 0 0 0	P	1 0 1 0 0 0 0	p	1 1 1 0 0 0 0
1	0 1 1 0 0 0 1	Q	1 0 1 0 0 0 1	q	1 1 1 0 0 0 1
2	0 1 1 0 0 1 0	R	1 0 1 0 0 1 0	r	1 1 1 0 0 1 0
3	0 1 1 0 0 1 1	S	1 0 1 0 0 1 1	s	1 1 1 0 0 1 1
4	0 1 1 0 1 0 0	T	1 0 1 0 1 0 0	t	1 1 1 0 1 0 0
5	0 1 1 0 1 0 1	U	1 0 1 0 1 0 1	u	1 1 1 0 1 0 1
6	0 1 1 0 1 1 0	V	1 0 1 0 1 1 0	v	1 1 1 0 1 1 0
7	0 1 1 0 1 1 1	W	1 0 1 0 1 1 1	w	1 1 1 0 1 1 1
8	0 1 1 1 0 0 0	X	1 0 1 1 0 0 0	x	1 1 1 1 0 0 0
9	0 1 1 1 0 0 1	Y	1 0 1 1 0 0 1	y	1 1 1 1 0 0 1
:	0 1 1 1 0 1 0	Z	1 0 1 1 0 1 0	z	1 1 1 1 0 1 0
;	0 1 1 1 0 1 1	[	1 0 1 1 0 1 1	{	1 1 1 1 0 1 1
<	0 1 1 1 1 0 0	\	1 0 1 1 1 0 0	|	1 1 1 1 1 0 0
=	0 1 1 1 1 0 1	]	1 0 1 1 1 0 1	}	1 1 1 1 1 0 1
>	0 1 1 1 1 1 0	^	1 0 1 1 1 1 0	~	1 1 1 1 1 1 0
?	0 1 1 1 1 1 1	—	1 0 1 1 1 1 1	delete	1 1 1 1 1 1 1

128 (2^7) code combinations for letters, numbers, and symbols. Table A3.1 lists the binary equivalents of ASCII codes. For instance, $S = 1010011$, while $s = 1110011$. Data files of ASCII codes can be created in one applications program and transferred to another applications program, where these data can be used for another function. This commonly occurs in finite element analysis, where the graphics are created in a CAD system and the coordinates are placed in data files and loaded into a finite element program for analysis.

WORD PROCESSING

Word processing is an odd term for a software that eases the tedium of writing. Computers were first created for data manipulation, or processing; later, as more sophisticated software was created, words could also be manipulated, or processed. The word processing software that is available today is quite sophisticated and easy to use. It can perform a variety of functions for you, but you must be the creative genius.

Word processing software is menu-deriven and can be used almost immediately, as the screen on which you see the typed copy has indexes to the help screens. Not only can you move quickly through the text and move chunks of text from one location to another, you can replace words easily. Many programs have a text replace feature, so a word can be replaced throughout the text, useful for correcting a misspelling. The software often includes a dictionary that will check the spelling of the text for you.

You can also import text from another document, so you do not have to rewrite repeated material. In addition you can determine how the output copy should look—words underlined, or double-struck for boldness.

For you as a student this software is very useful in all your courses, from English to engineering. The papers and reports that you write have the potential to reflect very well upon you. In engineering practice you will be communicating frequently, and word processing is an essential part of your job in report and proposal writing.

SPREADSHEETS

Spreadsheets are a computer application that has gained tremendous popularity with the accounting and financial world and is now gaining popularity with the engineering community. A spreadsheet is a computer program that analyzes data. Data analysis involves separating interrelated information into constituent parts, then varying the parts and determining the effect on the whole. The data must be separable and arranged in columns and rows.

The first personal computer spreadsheet, VisiCalc, was developed by Dan Bricklin and Bob Franklin to solve financial planning problems. The many similar programs today include Lotus 1-2-3 and Excel.

COMPUTER-AIDED DESIGN

The acronym CAD stands for computer-aided design. It usually appears in conjunction with CAM, computer-aided manufacturing. CAD/CAM refers not to one activity in the design or manufacturing process, but to many activities that are enhanced by using a computer, most often a computer workstation. The workstation is a very powerful computer that has accessories, such as display and input/output devices, and is connected with a mainframe computer. An engineer at the workstation can call up programs from the mainframe and use them in performing a design or analysis. The results can be seen visually at the workstation as well as communicated back to the mainframe. This same information can be transmitted from the mainframe to the computers used in the manufacturing process. CAD lets the engineer perform creative and multiple designs, as the computer does the necessary repetitive calculations.

Because of the workstation's computing power, memory, and storage capabilities, an engineer can see the results of a design and/or analysis on the CRT (cathode ray tube) display. The graphics display is most often in 1 to 16 colors or various shadings. The screen is divided in a matrix array. Each

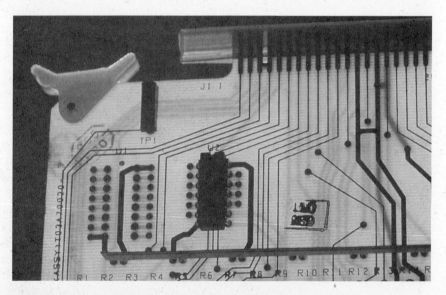

A design element circuit board. (Courtesy of Grumman Aerospace Corporation)

element, or cell, is called a pixel. The quality of the color display is determined by the number of pixels the computer can control. Some CRTs have a matrix of 4096 by 4096 pixels, but most have fewer elements; hence the resolution will not be as clear.

Using CAD systems for drawing requires a change in your orientation. When you draw a line between two points, you manually start at point x and draw the line to point y. Using a computer you define the points, x and y, and direct the computer to connect them with a straight, eliptical, or hyperbolic line. In addition the designer can assign line intensity, width, and coloration as well as other attributes.

Probably the CAD system you will first encounter in college is one using a wire-frame model. This is the simplest, most fundamental display of a three-dimensional model, where each edge appears as a line and the surfaces are transparent. Lines that are normally hidden are visible and must be removed by the designer. Some programs have hidden-line removal commands that create a quite realistic model. Figure A3.3 shows a three-dimensional model with and without hidden lines removed. Notice that the rendition (a), with all the lines included, is very difficult to visualize three-dimensionally. Removing the hidden lines, as in (b), helps significantly in the visualization process. While these models are very useful in drafting work and other two-dimensional applications, there are limitations in not having the third dimension. The computer cannot distinguish between inside and outside areas, and surface information is missing. Both are important to the manufacturing of three-dimensional objects.

The next level of CAD/CAM allows the surfaces to be modeled for certain objects, This technique works particularly well with thin materials, such as sheet metal, where the difference between the inner and outer surfaces is not significant. The surface model is generated by creating a wire-frame model and placing planes between the wire-frame edges. The techniques are sufficiently sophisticated to allow complicated curved surfaces, including fillets, to be modeled. Figure A3.3c illustrates the machine part with surfaces shown, the clearest rendition for us.

Solid modeling is the most sophisticated of graphics display; the part can be modeled with internal and external surfaces and any intermediate parts, all with different colorations. The quality of picture is excellent, with all surface nuances shown. These sophisticated workstations can light the object from different angles, producing very artistic effects. It is truly a blending of art and engineering.

Not only can the object be created, it can be analyzed using a finite element program. The finite element method is a numerical method of solving a variety of physical problems, from heat transfer to stress analysis. For most applications an exact mathematical solution to the equations modeling the problem is impossible, and approximate techniques, such as the finite element method, are employed. In this situation the total region, such as an airfoil (a thin plate) attached to a frame on an airplane, as shown in Figure A3.4, is

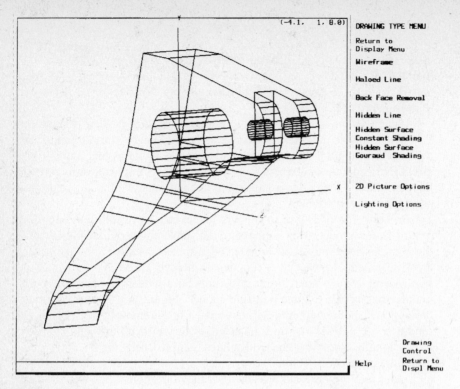

Figure A3.3(a)

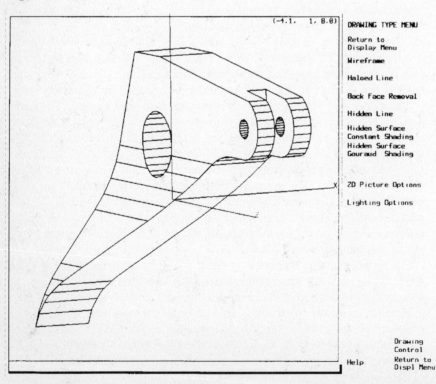

Figure A3.3(b)

Figure A3.3(c) (a) A wire-frame drawing of a machine part. (b) A wire-frame drawing of the same machine part, with the hidden lines removed. Notice the clarity you gain. (c) Solid modeling showing surfaces and shading of the same machine part.

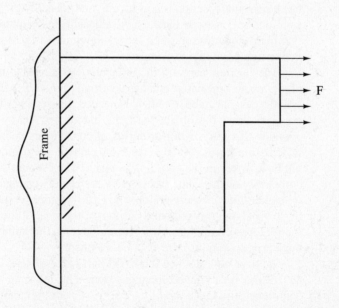

Figure A3.4 A thin uniform plate, attached to a frame and with a uniform force acting on it.

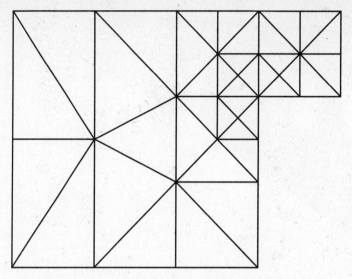

Figure A3.5 A finite element grid pattern for the plate in Figure A3.4. The element size varies as a function of change in stress gradient in the plate, with a smaller grid size in areas of high stress gradients.

divided into small elements as in Figure A3.5. The equations modeling the various physical parameters are written for each element, in simplified form because a small section is considered, and solved with the constraint that the solutions be consistent between the adjacent small elements. You can imagine that many equations must be solved simultaneously, requiring the processing power now available in computers.

The engineer can vary the element materials, their thickness, and method of joining in some cases, and can determine the stress at critical points. In this way the design can be modified as needed. In creating the design initially, the engineer drew the object with the aid of a CAD system. The computer contains this information in terms of coordinates defining the object. This information is transferred to the finite element program, so the engineer does not have to key in the coordinates again. The element size is dictated by the complexity of the shape where there are rapidly changing gradients in force or temperature. In these regions the grid size must be smaller than in regions of slowly changing gradients. Obviously, the smaller the elements the greater the required computing time and the greater the number of equations and boundary conditions that must be matched.

We have now reached the CAM part of CAD/CAM. The design is completed and needs to be manufactured. The information about the design component, its CAD data file, can be sent to the computer controlling the manufacturing process, often a manufacturing robot. This is currently atypical, because the CAD to CAM interface is still cumbersome.

The data from the CAD system must be compatible with the CAM system, and the manufacturer needs to invest in the robots and automated manufacturing systems that use the CAD in the first place. Not only must the company have the financial ability to fund such a conversion, but it requires a change, and decrease, in the work force. There are obvious social and political ramifications to such changes.

DATA BASES AND DATA-BASE PROCESSING

Information is key to making engineering and management decisions. Knowledge about new products, sales, manufacturing costs, inventory, parts, and personnel becomes important. Not only are these subjects individually important, but their interrelationships can be extremely valuable at times. Knowing the in-house inventory of parts may reduce the time required to undertake a new product line. Data-base technology allows associated data to be processed as a whole.

Rather than considering an industrial plant, let's consider an example involving your current environment, a university. Different areas of the university need related specific information each semester. For instance, information about the faculty teaching load is required to generate their paychecks, while some of this information is needed to schedule who is teaching what, when, and where. Related to the class schedules are the student data regarding who is attending what classes, so grades may be given. Figure A3.6 illustrates these three application files and the data that are required in them. Each file contains its own data, even though the same data will appear in more than one file.

Data-base problems occur when you want information that crosses file boundaries, such as the average salary of faculty teaching a certain course.

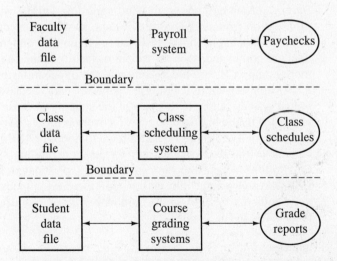

Figure A3.6 Three application files that are used in universities. There is no communication among the files.

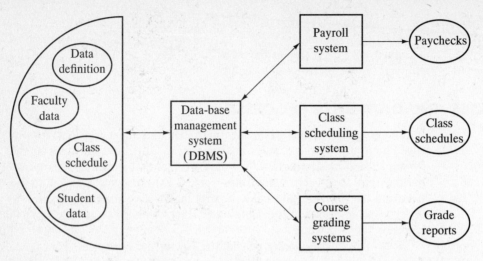

Figure A3.7 An integrated data-base management system allows information sharing. The boundaries between the data files are eliminated.

This requires information from the faculty file and the class data file. There is no certainty that the information in one file will be formatted in such a way as to be accessible by another file. Sometimes it is simply not worth the effort to get this information from the computer, unless the information is stored in a data base (Figure A3.7).

In this situation a data-base management system (DBMS) acts as system librarian. It stores the data and its description and retrieves this data for the applications programs as needed. The applications programs have not changed their function; they are just integrated into a whole by the DBMS. Several advantages spring to mind regarding DBMS: you can obtain more information from a given piece of data, as it may be correlated in a variety of ways; data duplication is reduced or eliminated; and inconsistencies in the data are eliminated, such as student or faculty name changes. The DBMS is more complex than a file system, and the applications programs need to be more sophisticated to handle the "generic" data. Because DBMS is an integrated system, failure in one part of the system can cause the entire system to break down, and in case of failure, recovery is more difficult. Backing up files is essential. The primary disadvantage to adopting DBMS is the conversion of data to the data-base system format and revising specific applications programs to accept the data in a more sophisticated form. However, the advantages of greater information gain are driving industry to adopt DBMS more fully, particularly as it can now be run on mainframes and minicomputers.

A3.3 COMPUTER PROGRAMMING LANGUAGES

When digital computers were first developed in the 1940s, and for about the following ten years, people had to communicate with the computer in machine language. Whereas this language is handy for computers, it is decidedly not easy to program and hence is error prone. Every computer manufacturer had its own machine language, and the computer instructions and memory addresses were specified in digits. Adding two values together might include these instructions:

```
150  58  410034
151  53  410038
152  50  421050
```

Machine languages, known as first-generation programming languages, were too difficult to use, and engineers soon developed a more symbolic language, assembly language programming, known as the second generation of languages. Each assembly language instruction is converted into machine language through the use of an assembler. To add two values in assembly language the program might include the following instructions:

```
LDA X    LOAD X IN ACCUMULATOR
ADD Y    ADD  Y TO ACCUMULATOR
STA Z    STORE ACCUM SUM AT Z
```

In this example X is loaded in the accumulator, Y is added to this value, and their sum, Z, is stored in a memory cell. This represents a tremendous advance over machine language programming, but it is still tedious. Enter the third-generation languages in the late 1950s and early 1960s, including BASIC, FORTRAN, and PL/1. In BASIC the summation problem is simply 100 Z = X + Y.

A printed circuit board after wave soldering. In wave soldering the circuit board is positioned close to the liquid solder surface in a tank. A small wave of solder is generated and passes across the circuit board, soldering the circuit connections. There is no trapped air or voids as would occur if the circuit board was placed on the liquid surface. (Courtesy of Grumman Aerospace Corporation)

In the 1980s fourth-generation languages are being developed and will continue to be created; they will be easier to use and more user-friendly, and will require less computer time to run. Table A3.2 shows the historical development of computer languages as well as their primary areas of application. The names of the languages are self-explanatory except for ADA, named for Lady Ada Lovelace, considered to be the world's first computer language developer in 1842. The computer was not electronic or electric, of course, but mechanical and known as Charles Babbage's analytic engine. The analytic engine could mechanically store 1000 numbers of 50 digits using ten position wheels. It was devised to store data and calculational results, perform mathematical operations, and input and output data. Lady Lovelace, the daughter of Lord Byron, devised a sequence of instructions for the engine to perform a set of complex calculations, the first computer program.

BASIC stands for Beginner's All-purpose Symbolic Instruction Code; it was developed at Dartmouth College in the 1960s. The word *beginner* is misleading, as it is a quite powerful language, very amenable to time-sharing systems and more importantly to personal computers. Not all computers use the same version of BASIC, but the versions are essentially the same. It is

Table A3.2 PRIMARY HIGH-LEVEL PROGRAMMING LANGUAGES

Language Name and Meaning	Year Introduced	Primary Application Areas
FORTRAN (*FOR*mula *TRAN*slation)	1957	Science and engineering
COBOL (*CO*mmon *B*usiness-*O*riented *L*anguage)	1959	Business data processing
ALGOL (*ALGO*rithmic *L*anguage)	1960	Science and engineering
APL (*A* *P*rogramming *L*anguage)	1962	Science and engineering (particularly time-sharing systems)
PL/1 (*P*rogramming *L*anguage/1)	1964	Business, science, and engineering
BASIC (*B*eginners *A*ll-purpose *S*ymbolic *I*nstruction *C*ode)	1965	Business, science, and engineering (particularly time-sharing systems)
Pascal (named after Blaise Pascal)	1971	Business, science, engineering, and education
C (the language after B)	1972	Scientific, engineering, and system software development
Ada (named after Augusta Ada Lovelace)	1980	Real-time, embedded computer systems

very user-friendly and can handle sophisticated circuit analysis and mechanical analysis problems.

FORTRAN is another language that you may encounter; it, Pascal, and BASIC form the core of current general-purpose engineering and scientific languages. FORTRAN is an acronym for "formula translation" and was developed with engineers and scientists in mind. During the time of its development key-punched cards were used, and its structure coordinates with this method of inputting data and formulae. Many FORTRAN programs are used in industry on mainframe computers, so the move from it to other languages will be slow. However, much of the program development for personal computers uses other languages, often BASIC, so a countervailing force is being created in the marketplace. This is not a great obstacle for you; when you learn one of these languages, switching to another is not difficult. The logic sequencing is the same.

Appendix Four

Algebraic and Trigonometric Problem Solving

One of the wonderful attributes of an engineering education is the ability to solve problems. However, one of the difficulties that many engineering students have is in the solution of algebraic word problems. Much of engineering involves the solution of just such problems, hence algebraic facility is necessary for the successful study and practice of engineering.

The function of this supplement is to present a variety of algebraic and trigonometric problems to enhance your problem solving abilities. This supplement cannot and does not replace a text in pre-calculus or algebra, but is intended to provide a quick review of algebra and trigonometry fundamentals. Examples and homework problems relate to and are drawn from various fields of engineering and science.

In addition, a problem solving methodology is used as shown below. This is not a "cookbook" approach and it may seem cumbersome to use on elementary problems, but following the approach can develop your confidence and ability to solve quite complicated problems.

Given: State in your own words what the given information is. This requires that you think about the problem satement and not simply restate it.

Find: State what must be determined.

Sketch and Given Data: Draw a sketch of the problem, including relevant data from the problem statement. The purpose here is to conceptualize what is happening and translate the word problem into sketches and diagrams. This is a key point in the problem-solving methodology.

Assumptions: List assumptions made in modeling the particular problem.

Analysis: Apply the assumptions to the governing equations and relationships. You should endeavor to work with the equations as long as possible before substituting numerical data in them. Substitute the data when the equations have been reduced to their final form.

LINEAR EQUATIONS

A linear equation is of the form $ax + b = 0$, where a and b are constants ($a \neq 0$) and x is the variable. The solution of the equation is $x = -b/a$, quite easy. The challenge is in translating a word problem into a linear equation, not so easy.

EXAMPLE 1

In surveying a rectangular lot a civil engineering student reports that one side is 4 meters longer than twice the other side. Find the lot's area and perimeter.

SOLUTION

Given: A rectangular lot and the relationship between the two sides.
Find: The lot's area and perimeter.
Sketch and Given Data:

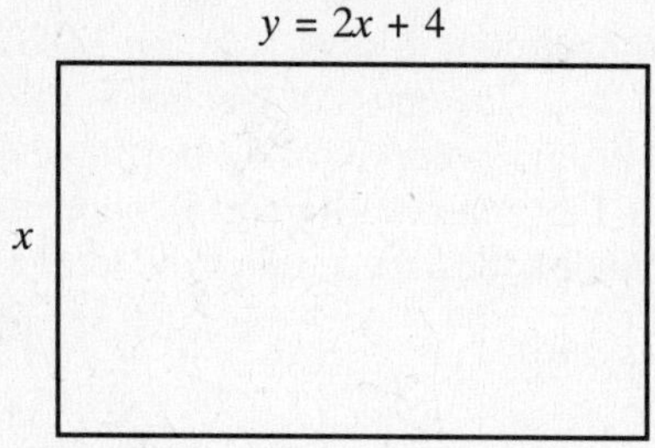

Figure A4.1

Assumptions: None.

Analysis: Let the shorter side be x as indicated in the sketch. The other side is y. Use the connection provided in the problem information to determine y's length.

$$y = 2x + 4 \text{ meters}$$

The perimeter, P, is

$$P = 2x + 2(2x + 4) = 6x + 8 \text{ meters}.$$

The area, A, is

$$A = (x)(2x + 4) = 2x^2 + 4x \text{ meters}^2.$$

EXAMPLE 2

In a chemical processing plant a mixture containing 30% alcohol must be obtained by adding a 60% alcohol solution to 40 kilograms of a 20% alcohol solution. Find the kilograms of the 60% alcohol solution required. All percentages are on a mass basis.

SOLUTION

Given: A known mass of an alcohol solution, the percent alcohol required and the percent alcohol of mixture to be added.

Find: The amount of a 60% alcohol solution that must be added to obtain the correct final solution.

Sketch and Given Data:

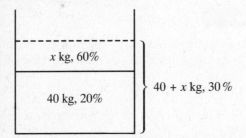

Figure A4.2

Assumptions: Conservation of mass of alcohol.

Analysis: Let x be the amount of 60% solution added. The total number of kilograms will be $40 + x$. The mass of alcohol in the mixture is the percent times the total mass. From assumption 1 we postulate that none of the alcohol can disappear. Thus the conservation of alcohol requires that the initial alcohol, plus the alcohol added must equal the total alcohol.

$$(0.20)(40) + (0.60)(x) = (0.30)(40 + x)$$
$$8 + 0.6x = 12 + 0.3x$$
$$x = 13.33 \text{ kilograms}$$

EXAMPLE 3

Three pipes may be used to fill a tank. A student measures the times to fill the tank for each of the pipes and records the following data: 20 minutes, 15 minutes and 40 minutes. Determine the time to fill the tank using all three pipes simultaneously.

SOLUTION

Given: A tank and the times to fill the tank individually from three pipes.
Find: The time to fill the tank using all three pipes simultaneously.
Sketch and Given Data:

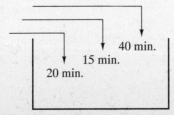

Figure A4.3

Assumptions: The flow through any pipe does not change when all are used simultaneously.

Analysis: Let the time to fill the tank be t. In one minute the amount of liquid entering the tank is

$$(1)(1/20 + 1/15 + 1/40).$$

Note that the first pipe fills 1/20th of the tank in a minute, the second pipe fills 1/15th of the tank in a minute and the third pipe fills 1/40th of the tank in a minute. In t minutes the tank will be filled, thus

$$(t)(1/20 + 1/15 + 1/40) = 1$$
$$t = 7.06 \text{ minutes.}$$

PROBLEMS

1. An isosceles triangle has a perimeter of 50 cm. The base, the shortest side, is one-half the length of the two longer sides. Determine the length of the sides.
2. A student received grades of 85, 84, and 91 on her first three chemistry quizzes. What must the grade of her fourth quiz be so she has a 90 average?
3. A student received grades of 78, 82 and 72 on three tests in physics. The final exam counts as two test grades. What must she score on the final exam to have an average of 80 for the course?
4. The octane rating of a gasoline is determined by comparing an engine's peak pressure from an actual gasoline to a standard value. The octane rating of a mixture is determined by the volumetric addition of the fuels, thus equal volumes of 80 octane and 100 octane gasoline yields a 90 octane mixture. Determine how many gallons of 95 octane fuel must be added to 100 gallons of 85 octane fuel to obain a mixture with an octane rating of 91.
5. The pumps at a service station blend 87 octane gasoline with 95 octane gasoline to obtain an octane rating in between the two. A customer receives 15 gallons of 92 octane gasoline. How many gallons of 87 octane fuel were used?
6. A tank holds 500 kilograms of brine with a salt concentration of 20% by mass. How much water must be evaporated so the concentration rises to 50%?
7. To raise money for their club, students hosted a party. They collected $1400 from 280 people who attended. Club members were charged $2.50 and non-members $6.00 for attendance. How many club members attended the party?
8. The radiator of an automobile holds 4 gallons of a 10% antifreeze/water mixture. The percentage of antifreeze must be raised to 25% by draining some of the mixture and adding 100% pure antifreeze. How much mixture must be drained? All percents are on a volume basis.

9. A power plant burns coal which should have an average sulfur value of not more than 1.5%. 100 tons of coal are available with 2.5% sulfur. In addition, there are supplies of coal with 0.75% and 1.1% sulfur. Determine the tons of each that must be added so the mixture contains 200 tons at 1.5% sulfur.
10. A ceramic clay contains 50% silica, 10% water and 40% other minerals. Determine the percentage of silica on a dry (water-free) basis.
11. Gold has a value of $12 a gram. A student finds a large gold ore nugget, containing gold and quartz, that weighs 1000 grams. The density of gold is 19.3 g/cm^3, the density of quartz is 2.5 g/cm^3 and the density of the nugget is 6.5 g/cm^3. What is the value of the gold in the nugget?
12. A tank may be filled using pipe A or B with times of 10 and 20 minutes, respectively. When pipe C is used simultaneously with pipes A and B, it takes 5 minutes to fill the tank. How long does it take to fill the tank using only pipe C?
13. Two workers can individually assemble a device in 3 hours and 5 hours, respectively. How long would it take to assemble the device if they worked together?
14. An airplane flies with a velocity of 250 mph when there is no wind. In flying with the wind it travels a certain distance in 4 hours. However, in flying against the wind it can only travel 60% of that distance. What is the wind's velocity?

SIMULTANEOUS LINEAR EQUATIONS

Consider the equation $ax + by = c$ where x and y are variables and a, b, and c are non-zero constants. This is a linear equation in two unknowns. If there are two equations such as

$$a_1x + b_1y = c_1$$
$$a_2x + b_2y = c_2$$

and they are independent of one another and consistent, then they may be solved simultaneously to determine the common value of x and y. This may be done by addition and subtraction, substitution or graphically. When the equations are plotted, their intersection provides the solution. Dependent equations form the same line and inconsistent equations form parallel lines with no point of intersection.

EXAMPLE 4

A manufacturer has in stock 10 packages of part X and 12 packages of part Y. The total cost for these units was $548. Additional supplies are purchased, 4 packages of part X and 3 packages of part Y for a total cost of $170. Determine the cost for part X and part Y.

SOLUTION

Given: The total cost and number of units for two parts, X and Y. Purchases were made on two separate occasions.

Find: The unit cost for X and Y.

Sketch and Given Data:

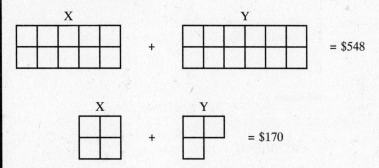

Figure A4.4

Assumptions: The unit price did not change.

Analysis: In this case we know the total cost, but not the individual cost of X and Y. Let

$$x = \text{cost of a unit of X}$$
$$y = \text{cost of a unit of Y.}$$

The total cost of the purchases was

$$10x + 12y = 548 \qquad \text{(a)}$$
$$4x + 3y = 170. \qquad \text{(b)}$$

Two methods of solution will be illustrated. To solve by substitution, solve Equation (b) for y.

$$y = \frac{170}{3} - \frac{4}{3}x \qquad \text{(c)}$$

Substitute this value into Equation (a).

$$10x + 12\left(\frac{170}{3} - \frac{4}{3}x\right) = 548$$
$$10x + 680 - 16x = 548$$
$$-6x = -132$$
$$x = 22$$

Substitute into either Equation (a), (b), or (c). Selecting (c),

$$y = \frac{170}{3} - \frac{4}{3}(22) = 27.33.$$

To solve by addition, multiply Equation (b) by 4 and subtract this from Equation (a).

$$10x + 12y = 548$$
$$-16x - 12y = -680$$
$$\overline{-6x = -132}$$
$$x = 22$$

Substitute back in either Equation (a) or (b) and solve for y. Substituting in Equation (a) yields

$$10(22) + 12y = 548$$
$$y = 27.33.$$

PROBLEMS

15. A chemical product uses two compounds, X and Y, in the ratio by weight of $X/Y = 1.5$. The total mass of product is 1000 kilograms. How many kilograms of each compound is required?
16. A computer manufacturer ships 200 computers to two different stores, A and B. It costs $4.50 to ship to A and $3.75 to ship to B. The total shipping invoice was $806.25. How many computers were shipped to each location?
17. Tank A contains a mixture of 100 liters of water and 50 liters of alcohol while tank B has 120 liters of water and 30 liters of alcohol. How many liters should be taken from the tanks to create an 80 liter mixture that is 25% alcohol by volume?
18. In a material science laboratory a 100 gram alloy is found to contain 20% copper and 5% tin by weight. How many grams of pure copper and pure tin must be added to this alloy to produce another alloy that is 30% copper and 10% tin?
19. Two objects move in a circular path of 276 feet at different but constant velocities. Assuming they do not collide with one another, when they start at the same point and move in the same direction they pass each other every 23 seconds. When they start at the same point and move in opposite directions, they pass each other every 6 seconds. Determine their velocities.

LINEAR SYSTEMS WITH THREE VARIABLES

It is possible to extend the technique of solving two simultaneous equations to three, or more, simultaneous equations. Usually when three or more simultaneous equations are solved, computers assist us in the process. However, there are occasions when it is necessary to perform the operations by hand

and one certainly should understand the manipulations the software is performing.

Consider the following set of equations.

$$2x + y - z = 2 \qquad \text{(a)}$$
$$x + 3y + 2z = 1 \qquad \text{(b)}$$
$$x + y + z = 2 \qquad \text{(c)}$$

The values of x, y, and z that simultaneously satisfy the equations may be found by addition and substitution. The next section on matrices will discuss a more general method of solution. In this case, first add Equations (a) and (c), eliminating the variable z, yielding

$$3x + 2y = 4. \qquad \text{(d)}$$

Eliminate z from another set of two equations, say (a) and (b), by multiplying (a) by 2 and adding the equations together which yields

$$\begin{array}{c} 5x + 5y = 5 \\ x + y = 1. \end{array} \qquad \text{(e)}$$

Solve Equations (d) and (e) simultaneously, substituting x in terms of y from Equation (e) into Equation (d).

$$3(1 - y) + 2y = 4$$
$$y = -1$$

From Equation (e), then

$$x = 1 - y = 1 - (-1) = 2.$$

Solve for z by substituting into one of the original equations. Pick Equation (c), as that is the simplest.

$$2 - 1 + z = 2$$
$$z = 1$$

Thus the simultaneous solution of the three independent equations is $x = 2$, $y = -1$, $z = 1$. An important habit to develop is verifying the results. Substitute the answers back into the original equation, making sure they satisfy the problem. This serves as a "double check" on your calculations.

There are many applications of simultaneous equations in engineering, so facility in developing the equations as well as in solving them is very important.

EXAMPLE 5

An agricultual engineer is developing a new animal feed which should have 22 kilograms of protein, 28 kilograms of fat and 18 kilograms of fiber. The engineer has available three types of plants: corn, cottonseed and soybeans. The table below indicates the kilograms of protein, fat and fiber per kilogram of plant food. The percent totals do not reach unity as other nutrients are contained in the plant.

	Corn	Cottonseed	Soybeans	Total (kg)
Protein/kg	0.25	0.2	0.4	22
Fat/kg	0.4	0.3	0.2	28
Fiber/kg	0.3	0.1	0.2	18

SOLUTION

Given: The requirements for an animal feed in terms in terms of protein, fat and fiber as well as the percentages that these components make up in the three available plant foods that will be used to make the animal feed.

Find: The kilograms of each plant food that must be used to make the animal feed.

Sketch and Given Data:

| Corn
25% protein
40% fat
30% fiber
x | Cottonseed
20% protein
30% fat
10% fiber
y | Soybean
40% protein
20% fat
20% fiber
z | Total
22 kg
28 kg
18 kg |

Figure A4.5

Assumptions: None.

Analysis: Let x = kilograms of corn, y = kilograms of cottonseed and z = kilograms of soybeans required. The 22 kilograms of protein must be made by combining

$$0.25x + 0.2y + 0.4z = 22.$$

The 28 kilograms of fat is found by combining

$$0.4x + 0.3y + 0.2z = 28.$$

The 18 kilograms of fiber is found by combining

$$0.3x + 0.1y + 0.2z = 18.$$

Write the equations in whole numbers by multiplying the equations by 100 which yields

$$25x + 20y + 40z = 2200 \qquad \text{(a)}$$
$$40x + 30y + 20z = 2800 \qquad \text{(b)}$$
$$30x + 10y + 20z = 1800. \qquad \text{(c)}$$

Subtract Equation (c) from Equation (b) which yields

$$10x + 20y = 1000$$
$$x + 2y = 100. \qquad \text{(d)}$$

Multiply Equation (b) by two and subtract Equation (a) from it which yields
$$55x + 40y = 3400. \tag{e}$$

Solve Equation (d) for x in terms of y and substitute into Equation (e), which yields
$$55(100 - 2y) + 40y = 3400$$
$$-70y = -2100$$
$$y = 30 \text{ kilograms.}$$

From Equation (d)
$$x = 40 \text{ kilograms}$$

and from Equation (c)
$$30(40) + 10(30) + 20z = 1800$$
$$z = 15 \text{ kilograms.}$$

Thus the feed should contain 40 kilograms of corn, 30 kilograms of cottonseed and 15 kilograms of soybeans.

MATRIX SOLUTION OF LINEAR SYSTEMS

There are several ways that matrices may be used to solve linear systems. What follows will describe one method. Consider a set of three equations and three unknowns:

$$a_1 x + b_1 y + c_1 z = d_1$$
$$a_2 x + b_2 y + c_2 z = d_2$$
$$a_3 x + b_3 y + c_3 z = d_3.$$

This set of equations may be written in an abbreviated form as

	1	2	3	4
Row 1	a_1	b_1	c_1	d_1
Row 2	a_2	b_2	c_2	d_2
Row 3	a_3	b_3	c_3	d_3

Columns.

This rectangular array of numbers is called a matrix. Each number is called an element in the array and is identified by its row and column number; thus, the number b_1 is in element (1, 2). The last column is separated from the first three by a vertical line; this indicates that it augments the matrix formed by the equations' coefficients.

Consider the following set of equations

$$x - 2y + z = 5$$
$$-2x + 4y - 2z = 2$$
$$2x + y - z = 2.$$

The augmented matrix is

$$\begin{array}{ccc|c} 1 & -2 & 1 & 5 \\ -2 & 4 & -2 & 2 \\ 2 & 1 & -1 & 2 \end{array}.$$

For any augmented matrix of a set of linear equations, the following transformations result in an equivalent matrix:

1. Any two rows may be interchanged;
2. The elements of any row may be multiplied by a nonzero, real number;
3. Any row may be changed by adding to its elements a multiple of the corresponding elements of another row.

We will use the Gauss-Jordan method to reduce the augmented system matrix to a unity matrix of the form

$$\begin{array}{ccc|c} 1 & 0 & 0 & a \\ 0 & 1 & 0 & b \\ 0 & 0 & 1 & c \end{array}$$

where a, b, and c become the solutions of x, y, and z that solve the set of simultaneous equations.

Consider first a linear system of two equations

$$3x - 4y = 1$$
$$5x + 2y = 19.$$

The augmented matrix is

$$\begin{array}{cc|c} 3 & -4 & 1 \\ 5 & 2 & 19 \end{array}.$$

Multiply each element in row 1 by 1/3 to get a one in element (1, 1).

$$\begin{array}{cc|c} 1 & -4/3 & 1/3 \\ 5 & 2 & 19 \end{array}$$

To obtain a zero in element (2, 1), multiply each element of row one by -5 and add it to the element in row two.

$$\begin{array}{cc|c} 1 & -4/3 & 1/3 \\ 0 & 26/3 & 52/3 \end{array}$$

To obtain a one in element (2, 1), multiply each element in row two by 3/26.

$$\begin{array}{cc|c} 1 & -4/3 & 1/3 \\ 0 & 1 & 2 \end{array}$$

To obtain a zero in element (1, 2), multiply each element of row two by 4/3 and add this value to the corresponding element in row one.

$$\begin{array}{cc|c} 1 & 0 & 3 \\ 0 & 1 & 2 \end{array}$$

Thus the solution to the two original equations is $x = 3$, $y = 2$. This may be extended to a system with three equations. Obviously, it is time consuming, but it is procedural and as long as one is careful, the solution to sets of equations may be determined. Very often students will have spreadsheets or other equation solvers available which have built-in matrix equation solvers.

PROBLEMS

20. Solve Example 5 using matrix methods.
21. A service station sells three grades of gasoline, regular, premium and super. One day the station sold 150 gallons of regular, 400 gallons of premium and 130 gallons of super for a total of $909. The next day it sold 170 gallons of regular, 380 gallons of premium and 150 gallons of super for $931. The price difference per gallon between super and regular is one-half the difference between premium and regular. Determine the cost per gallon for each grade of gasoline.
22. A chemical engineer has three salt solutions available, 5%, 15% and 25%, to make 50 liters of a 20% saline solution. There is much more 5% solution available, so a requirement is to use twice as much 5% solution as the 15% solution. Determine the amount of each salt solution that is used to make the mixture.
23. A manufacturing company produces two products, I and II, that require time on machines A and B. Product I requires 1 hour on A and 2 hours on B, while product II requires 3 hours on A and 1 hour on B. The company is open 16 hours a day, with the machines operating 15 hours per day. What is the number of each product that can be produced daily?
24. Three grades of resin are available which may be mixed together to form a fourth resin. The costs of the initial resins are $4.60, $5.75, and $6.50 per pound. The mixture value will be $5.25 per pound and 20 pounds are needed. In addition, the amount of the least expensive resin should be equal to the total amount of the other two. Determine the amount of each resin needed.
25. An engineering club holds a benefit party and collects a total of $2480 consisting of five, ten and twenty dollar bills. The total number of bills is 290. The value of the total number of tens is $60 more than the value of the total number of twenties. Determine the number of each type of bill the club has.

QUADRATIC EQUATIONS WITH ONE UNKNOWN

A quadratic equation has the form $ax^2 + bx + c = 0$ where x is the variable and a, b, and c are constants, a not equal to zero. There are several methods used for solving this equation, such as factoring, completing the square, graphical and using the quadratic formula. The quadratic formula is

$$x = \frac{-b \pm \sqrt{b^2 - 4ac}}{2a}.$$

There will be two roots to a quadratic equation; that is, two values of x will satisfy the equation. Often only one is physically possible, and this is the value selected. For instance, given the equation

$$x^2 + 3x - 4 = 0$$

the roots may be found by factoring, $(x + 4)(x - 1)$, or $x = -4$ or $= +1$. Applying the quadratic formula yields the same result,

$$x = \frac{-3 \pm \sqrt{9 - 4(1)(-4)}}{2(1)} = \frac{-3 \pm 5}{2} = 1 \quad \text{or} \quad -4.$$

EXAMPLE 6

A student is given information about a rectangular room which can be used for a club activity. The room's perimeter is 52 feet and the area is 160 square feet. Determine the room's dimensions.

SOLUTION

Given: The area and perimeter of a rectangular room.
Find: The room's dimensions.
Sketch and Given Data:

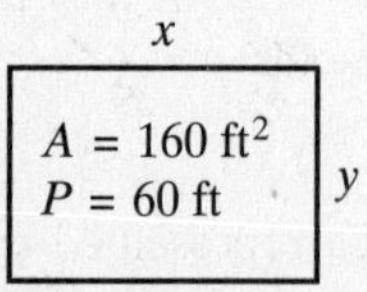

x

$A = 160 \text{ ft}^2$
$P = 60 \text{ ft}$ y

Figure A4.6

Assumptions: none
Analysis: The perimeter is the sum of the sides, hence

$$P = 52 = 2x + 2y$$
$$x + y = 26. \tag{a}$$

The area is

$$A = 160 = xy. \tag{b}$$

Solve Equation (a) for y in terms of x ($y = 26 - x$) and substitute into Equation (b).

$$160 = x(26 - x)$$
$$x^2 - 26x + 160 = 0$$
$$(x - 10)(x - 16) = 0$$
$$x = 10, 16$$

The dimensions of the room are 10 feet and 16 feet.

EXAMPLE 7

A supplier bought a certain number of copies of software for $1800 and sold all but six copies. The supplier made a profit of $20 per copy for each package of software sold. The supplier decided to reinvest the revenue in more copies and purchased 30 more copies than previously with the money. Find the unit cost of the software.

SOLUTION

Given: The total cost for an unknown amount of software, the profit for the copies sold and the number of additional copies that can be purchased.

Find: The unit cost of the software.

Sketch and Given Data:

```
total cost $1800
x = cost per copy
```
Figure A4.7

Assumptions: The unit cost of the software did not change.

Analysis: Let x be the cost per copy of software purchased for $1800. The total number of copies is $1800/x$. The total revenue received is

(number of copies sold)*(income per copy) = total revenue.
$$(1800/x - 6)*(x + 20) = TR. \qquad \text{(a)}$$

This value of total revenue may also be represented by the total number of new copies of software purchased.

(number of copies purchased)*(cost per copy) = total revenue.
$$(1800/x + 30)*(x) = TR. \qquad \text{(b)}$$

Equate Equations (a) and (b).

$$(1800/x - 6)(x + 20) = (x)(1800/x + 30)$$
$$36x^2 + 120x - 36{,}000 = 0$$
$$x^2 + 3.333x - 1000 = 0$$

Use the quadratic formula to solve for x.

$$x = \frac{-3.333 + -\sqrt{11.111 - (4)(1)(-1000)}}{(2)(1)}$$

$$x = 30, -33.33$$

The only physically possible answer is $30 per copy.

PROBLEMS

26. Two engineers are working on a project and complete it in 10 days. When each works alone, it takes the first engineer 5 days more than the second to complete the project. Determine the time it takes each, individually, to complete the project.

27. A right triangle is formed from a wire 60 cm long. The triangle's hypotenuse is 25 cm. Find the length of the other sides.

28. A student is given a 9 inch by 12 inch piece of paper and is to construct an open box by cutting equal squares from each of the corners of the paper and then folding up the sides. The base area should be 60 square inches. Find the length of the sides of the squares that are removed.

29. A student is driving home, a distance of 150 miles, for the weekend. From previous experience, the student knows that increasing the average speed 10 miles/hour could reduce the time of the trip by 35 minutes. What is the actual average speed?

30. A supplier purchased an unknown number of computer monitors for a total cost of $7200. Each was sold for a price of $400. The supplier received a total profit equal to the cost of only 8 monitors. Determine the number of monitors originally purchased.

31. Two pipes, I and II, can be used to fill a tank. Pipe I fills the tank in four hours. If pipe II is used by itself, it takes 3 hours longer than if both pipes are used simultaneously. Determine the time it takes to fill the tank with pipe II.

EXPONENTIAL AND LOGARITHMIC FUNCTIONS

An exponential function may be defined as $f(x) = a^x$. Examples of this are: $a^{2.5}$, 2^x, and 3^{-1}. Exponential growth is frequently encountered in engineering, science and mathematics. Imagine that a population is increasing at a given percentage per year, such as 6%. If the initial population is I_0 then the population after n years is $I = I_0(1 + .06)^n$. If a certain animal species increases at a rate of 6% and its initial population is 10,000, then the population after 5 years is $I = (10,000)(1.06)^5 = 13,382$.

A very useful exponential function occurs when $f(x) = e^x$, where e is an irrational number, 2.718281828. The exponential function is used to model many natural systems, such as the exponential decay of radiation, transient responses in electrical and mechanical systems, and population change.

A logarithm is the exponent of an exponential function that represents a real, positive number. There are two logarithms that are commonly used, logs to the base ten (common logarithms) and logs to the base e (natural logarithms). Consider the number 820, represented as 10^x where x is 2.9138 or e^x where x is 6.7093. The exponent is the logarithm of the number to a given base. Thus,

$$\log_{10} 820 = 2.9138$$
$$\log_e 820 = \ln 820 = 6.7093.$$

An example of a log function commonly used is the decibel rating of sound. The loudness of sound is measured in decibels (db) and is a $\log_{10}$ equation.

$$db = 10 \log_{10} I/I_0$$

where I_0 is the intensity of the faintest sound that can be heard and I is the intensity of an actual sound. If $I = 10{,}000 I_0$, then

$$db = 10 \log_{10} [10{,}000 I_0/I_0] = 40.$$

For the db to increase from 40 to 50, the intensity must increase from $10{,}000 I_0$ to $100{,}000 I_0$.

Sometimes students have difficulty in solving exponential and logarithmic equations. Consider the following equation

$$50 = 1000(1 - e^{-3t})$$

and solve it for t. The following steps are used:

$$0.05 = 1 - e^{-3t}$$
$$0.95 = e^{-3t}$$
$$\ln(0.95) = -0.05129 = -3t \ln(e) = -3t(1)$$
$$t = 0.017.$$

Solve the following equation for x.

$$y = (a)[\ln(1 + x/a)]$$
$$y/a = \ln(1 + x/a)$$
$$e^{y/a} = 1 + x/a$$
$$x = a(e^{y/a} - 1)$$

PROBLEMS

32. Assume that the annual rate of inflation is 5%. Determine how long it will take for prices to double.

33. The half-life of radioactive carbon 14 is 5700 years. After a plant or animal dies, the level of carbon 14 decreases as the radioactive carbon 14 disintegrates. The decay of radioactive material is given by the relationship, $A = a_0 e^{-kt}$, where A_0 is the initial amount of material at time zero and t represents the present time measured from time zero in years. For carbon 14, $k = 1.216 \times 10^{-4}$ years^{-1}. Samples from an Egyptian mummy show that the carbon 14 level is one-third that found in the atmosphere. Determine the approximate age of the mummy.

34. Paint from cave drawings in France indicate a carbon 14 level of 15% that of carbon found in the atmosphere. Determine the aproximate age of the drawings.
35. The amount of a certain chemical, A, that will dissolve in solution varies exponentially with the Celsius temperature, T, according to the equation $A = 10e^{0.01T}$. Determine the temperature that allows 15 grams of chemical to be dissolved.
36. Newton's law of cooling describes the cooling or heating of an object by a fluid (liquid or gas). The temperature variation with time is given by the equation $T(t) = T_0 + A\,e^{-kt}$ where T_0 is the surrounding fluid temperatue, A is a constant equal to 100, k is a constant equal to 0.1 and t is the time in minutes. Determine the time it will take a cup of hot coffee to cool to 30°C in a room at 20°C.

TRIGONOMETRY

Algebra and trigonometry are related in that trigonometric problems often require algebraic problem solving ability. Figure A4.8 illustrates a right triangle. The trigonometric functions are:

Trigonometric Ratio	Abbreviation	Definition	
sine of θ	$\sin \theta$	$\dfrac{\text{side opposite}}{\text{hypotenuse}}$	$= \dfrac{a}{c}$
cosine of θ	$\cos \theta$	$\dfrac{\text{side adjacent}}{\text{hypotenuse}}$	$= \dfrac{b}{c}$
tangent of θ	$\tan \theta$	$\dfrac{\text{side opposite}}{\text{side adjacent}}$	$= \dfrac{a}{b}$
cotangent of θ	$\cot \theta$	$\dfrac{\text{side adjacent}}{\text{side opposite}}$	$= \dfrac{b}{a}$
secant of θ	$\sec \theta$	$\dfrac{\text{hypotenuse}}{\text{side adjacent}}$	$= \dfrac{c}{b}$
cosecant of θ	$\csc \theta$	$\dfrac{\text{hypotenuse}}{\text{side opposite}}$	$= \dfrac{c}{a}$

The Phythagorean theorem is also used extensively in the solution of right triangles, $c^2 = a^2 + b^2$.

Angles are measured either in degrees or radians. There are 360 degrees in a circle and 2π radians. Figure A4.9 illustrates angles using degree measurements and Figure A4.10 degree measurements using radians. Note that one radian is the angle whose subtended arc has a length equal to the radius of the circle. Each degree is equal to $\pi/180$ radians. Calculators use either radians or degrees, so be sure to check which is required. In addition, computer programs often use the trigonometric angles measured in radians.

ALGEBRAIC AND TRIGONOMETRIC PROBLEM SOLVING

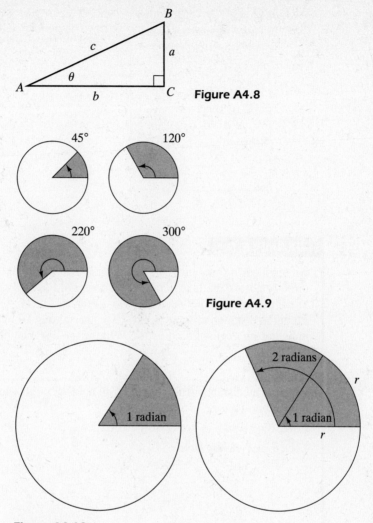

Figure A4.8

Figure A4.9

Figure A4.10

Convert 30 degrees into radians.

$$(30 \text{ degrees})(\pi/180 \text{ radians/degree}) = \pi/6 \text{ radians}$$

Convert 4 radians into degrees.

$$(4 \text{ radians})(180/\pi \text{ degrees/radian}) = 229.1 \text{ degrees}$$

PROBLEMS

37. A plot of land is a 270 degree sector with a 10 foot radius. Determine the area in square feet.

38. A curve along a highway is a circular arc 50 m long with a radius of curvature of 250 m. How many degrees does the highway change its direction along the arc?

39. Find the area of sector inside the square *ABCD*.

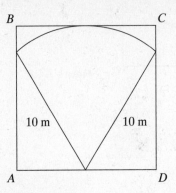

EXAMPLE 8

An engineering student visits a cottage located on a lake. Another cottage is directly across the lake. The owner does not have a boat and would like to know the distance between the two cottages. The engineering student measures the distance to point *A* as 310 m. A sighting from point *A* to cottage *B* indicates an angle of 55 degrees. What is the distance between the cottages?

SOLUTION

Given: The distance from cottage *C* to another point and the angle from that point to cottage *B*.

Find: The distance between the cottages.

Sketch and Given Data:

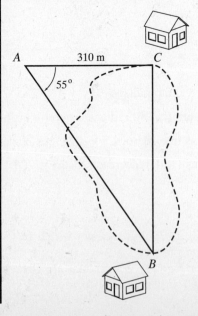

Figure A4.11

Assumptions: None.

Analysis: The distance that is required is line *CB*. From the definition of a tangent of an angle,

$$\tan A = \frac{CB}{AC}$$

$$\tan(55) = \frac{CB}{310}$$

$$CB = (310)(1.428) = 442.7 \text{ m}$$

PROBLEMS

40. The angle of elevation to the top of a flagpole is 40 degrees from a point 30 m from the base of the pole. What is the height of the pole?
41. A kite string forms an angle of 42 degrees with the ground when the entire 800 feet of string is used. What is the kite's elevation?
42. An 80-foot pole is stabilized by guide wires which run from the top of the pole to the ground. The wires are located 15 feet from the base of the pole. What length guide wire is required? What is the angle the wire makes with the ground?
43. A surveyor measures the angle of elevation of a mountain from point *A* and finds it to be 23 degrees. The surveyor moves 1/4 mile closer to the mountain and finds the angle of elevation to be 43 degrees. What is the height of the mountain? (Assume that both points and the base of the mountain are on the same line.)
44. A 12-foot flagpole stands at the edge of a building's roof. The angle of elevation from the ground 65 feet from the building to the top of the flagpole is 78 degrees. Determine the building height.

THE LAWS OF COSINES AND SINES

In general a triangle is not a right triangle. There are two laws that assist in providing relationships between sides and angles. Referring to Figure A4.12, the Law of Cosines is

$$A^2 = B^2 + C^2 - 2BC \cos \alpha.$$

Figure A4.12

The Law of Sines is

$$\frac{A}{\sin \alpha} = \frac{B}{\sin \beta} = \frac{C}{\sin \gamma}.$$

PROBLEMS

45. A diagonal of a parallelogram has a length of 60 inches and makes an angle of 20 degrees with one of the sides. The side has a length of 25 inches. Determine the length of the other side of the parallelogram.
46. Determine the length of AB.

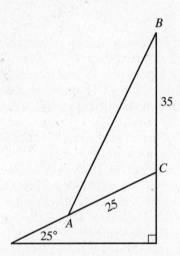

47. A surveyor is determining the distance between two points A and B located on a shoreline. The surveyor is located at point C and measures the distance AC to be 180 m and BC to be 120 m. The angle at C is 56 degrees. Find the distance AB.
48. In problem 47, if the angle at C is 130 degrees, find AB.
49. Two guide wires are attached to the top of a pole and are anchored into the ground on opposite sides of the pole at points A and B. The ground is the same elevation relative to the pole in all directions. The distance AB is 40 m and the angles of elevation at A and B are 70 and 55 degrees respectively. Determine the guy wire lengths.
50. An airplane is flying in a straight line and at constant elevation towards an airfield. At a given instant the angle of depression between the plane and airfield is 32 degrees. After flying two miles, the angle of depression is 74 degrees. What is the distance between the airplane and the airfield at the second point?

51. In the figure shown, points A and B are on the same side of a river where the distance AB is 600 feet. Determine the distance CD on the opposite side of the river.

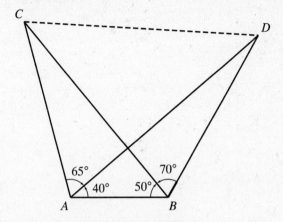

CIRCULAR SINE AND COSINE FUNCTIONS

Consider the function $y = \sin \theta$. For this function y varies between $+1$ and -1. Figure A4.13 illustrates a unit circle with various values for y as a function of $\sin \theta$. The function $y = \sin \theta$ may be plotted horizontally, the

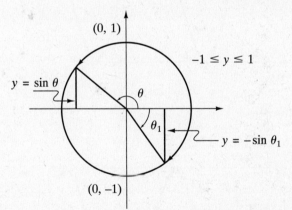

Figure A4.13

circle is "unrolled" as indicated in Figure A4.14. The sine function repeats itself every 2π radians, thus the sine function is periodic with a period of 2π. The entire unrolled function is shown in Figure A4.15.

In a similar fashion the cosine function may be plotted as shown in Figure A4.16. The amplitude of the function, $y = a \sin \theta$ is determined by

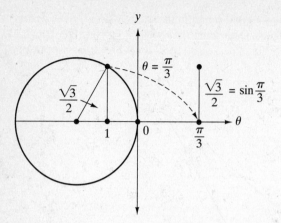

Figure A4.14

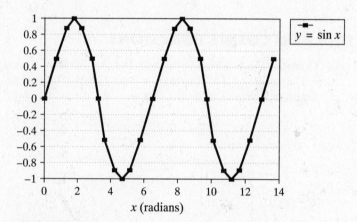

Figure A4.15

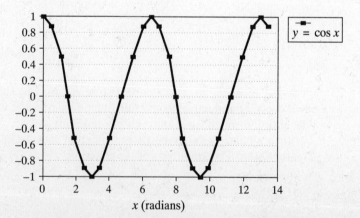

Figure A4.16

the value of a. In addition to plotting the functions separately, frequently one must plot combined functions, such as where x varies between 0 and 2π. In this situation plot each of the functions separately and then add the values together, plotting the composite function. Use of these functions in a spreadsheet simplifies this plotting. The result for the function $y = 2 \sin x + 3 \cos x$ is shown in Figure A4.17.

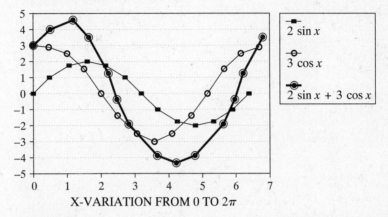

Figure A4.17

PROBLEMS

52. Sketch the following curves between $0 < x < 2\pi$.
   a) $y = \cos 2x$
   b) $y = -\sin 2x$
   c) $y = -5 \cos x$
   d) $y = 4 \sin (0.2x)$

53. Sketch the following functions for $0 < x < 2\pi$.
   a) $y = \sin x - \cos x$
   b) $y = 2 \cos x + \cos 2x$
   c) $y = \sin 2x + 0.5 \cos x$

Index

A

Absolute zero, 196
Abstract, 119
Accreditation Board for Engineering and Technology (ABET), 51
Accuracy, 179, 194
Aeronautical engineering, 52
Aerospace engineering, 52
Agricultural engineering, 53
Agricultural society, 22
Algebraic problems, 268
American Association of Engineering Societies (AAES), 48
American Institute of Electrical Engineers, 49
American National Standards Institute (ANSI), 95
American Society of Civil Engineers, 39, 49
American Society of Mechanical Engineers, 49
Analysis, 188, 219
Aristotle, 31
Art, 3
ASCII, 256
Automotive engineering, 54
Average, arithmetic, 167

B

Bacon, Francis, 24
Bar chart, 141
Bioengineering, 55
Blueprint, 136
Bose, Jagadis Chandra, 36
Building codes, 8

C

Chemical engineering, 56
Circle diagram, 140
Civil engineering, 58
Code of ethics (NSPE), 233
Communication, 116
  graphical, 136
  oral, 124
  written, 117
Computer
  applications, 255
  data bases, 263
  design use, 258
  operation of, 250
  processing cycle, 250
  programming languages, 264
  spreadsheets, 257
  word processing, 257
Computer-aided design (CAD), 258

Computer engineering, 59
Computer science, 59
Concurrent engineering, 69, 80
Conflict of interest, 108
Construction, 72
Consulting, 75
Conversion factor, 203
Conversions, unit, table of, 241
Craftsperson, 46
Creativity, 3, 216
Curriculum, 11
Curve-fitting, 177

D

Data-bases, 263
da Vinci, Leonardo, 34
Dark Ages, 33
DC-10 case, 96
Decision analysis, 204
   multicriteria, 206
   tree, 205
Descartes, René, 24
Design process, 71, 214
   additional considerations, 224
   creating solutions, 216
   problem definition, 215
   refinement and analysis, 219
   resolution and implementation, 221
Development, 70

E

Egypt, 25
   Great Pyramid of, 27
Electrical engineering, 60

Electronic engineering, 60
Encyclopedias, technical, 129
Engineer, root word, 34
Engineering, 2
   aeronautical, 52
   aerospace, 52
   agricultural, 53
   architectural, 54
   automotive, 54
   biomedical, 55
   ceramic, 55
   chemical, 56
   civil, 58
   computer, 59
   concurrent, 69, 80
   construction, 72
   consulting, 75
   creativity, 3
   curriculum, 11
   definition, 2
   design, 71
   development, 70
   electrical, 60
   electronic, 60
   environmental, 62
   ethics, 92
   geological, 67
   history of, 22
   industrial, 62
   management, 74
   manufacturing, 63, 72
   marine, 64
   materials, 65
   mechanical, 66
   memo, 124
   metallurgical, 65
   military, 34
   mining, 67
   nuclear, 67

ocean, 64
operations and maintenance, 73
petroleum, 68
plant, 73
quality assurance, 76
research, 69
sales, 74
societies, 48
teaching, 75
team, 69
Engineering Accreditation Commission (EAC), 51
Engineering Manpower Commission (EMC), 51
English Engineering Unit System, 197
Environmental engineering, 62
Ergonomics, 220
Error analysis, 179
Ethical problems, 92
Ethics, 92
  code of, 92, 233
Exponential functions, 282

F

Factor of safety, 8
Fields of engineering, 51
Finite element analysis (FEA), 262
Force, 197
Founder societies, 48

G

Gantt chart, 207
Geological engineering, 67
Graphical analysis, 136

Graphics, sketching, 154
Graphs, 142
Great Pyramid, 27
Great Wall of China, 27
Greek civilization, 29
Gutenberg, Johannes, 34

H

Handbooks, 129
Happiness, 101
Histogram, 173
Humanities, 11
Hyatt Regency, 101
Hydrogen bomb, 89

I

Indexes, 129
Industrial engineering, 62
Industrial revolution, 34
Invention, 225

J

"Just in time" manufacturing, 83

K

Kelvin, definition of, 200
Kilogram, definition of, 200

L

Library usage, 127
Line graphs, 142
Linear equations, 269

Linear regression analysis, 177
Linear systems, 275
Log-log graphs, 149
Logarithmic functions, 282

M

Management, 74
Manufacturing, 72
Manufacturing engineering, 63, 72
Marine engineering, 64
Mass, 195
Materials engineering, 65
Materials science, 65
Mathematical models, 190
Matrix solution, 277
Mechanical engineering, 66
Membership, engineering society, 50
Memo, 124
Mesopotamia, 25
Metallurgical engineering, 65
Metallurgy, beginnings of, 27
Microprocessors, 253
Middle Ages, 33
Mining engineering, 67
Moral dilemmas, 92

N

National Academy of Engineering (NAE), 48
National Society of Professional Engineers (NSPE), 92
Newton (unit of force), 201

NIMBY, 6
Note taking, 14
Nuclear engineering, 67
Nylon, invention of, 37

O

Ocean engineering, 64
Operations and maintenance, 73
Operating systems, 252
Oral communications, 124

P

Petroleum engineering, 68
Pictograph, 138
Political implications of technology, 5
Pounds force (lbf), 197
Pounds mass (lbm), 197
Precision, 179
Printing press, invention of, 34
Probability, 102, 171
Problem
    definition, 215
format of, 188
solving, 190
Products
    quality of, 94
    safety of, 94
Profession, 90
Professional organizations, 92
Professional registration, 91
Programming languages, 265
Public safety, 94

Q

Quadratic equation, 280
Quality, 76
Quality assurance, 76
Quality control, 76

R

RAM, 254
Range, 143
Reference sources, 128
Refinement and analysis, 219
Reliability, 10
Reports
  oral, 125
  written, 119
Research, 69
Resolution and implementation, 221
Résumé, 121
Risk assessment, 102
ROM, 253
Roman civilization, 30

S

Safety, 102
Sales engineering, 74
Scale, 143
Scientific notation, 193
Scientist, 46
Scratch plow, 26
Semilog graphs, 148
Significant figures, 193
SI metrics (Système International)
  decimalization, 199
  derived units, table of, 202
  fundamental units, table of, 200
  prefixes, table of, 201
  rules, 203
Simultaneous linear equations, 273
Sketching techniques, 154
Slope, 146
Spreadsheets, 257
Statistics, 166
  arithmetic mean, 167
  frequency distribution, 170
  median, 167
  mode, 168
  normal distribution, 170
  range, 169
  standard deviation, 169
Steam engine, 35
Steam pump, 35
Study habits, 14
Sturgis rule, 173

T

Tacoma Narrows Bridge, 103
Teaching, 75
Team, 46, 69
Technical reports, 118
  abstract, 119
  conclusions and recommendations, 120
  organization of, 118
  references, 121
  title page, 119
Technician, 46

Technological team, 46
Technologist, 46
Technology, 4
   creation of, 4
   impact on society, 5
   political implications of, 4
Technology Accreditation Commission (TAC), 51
Temperature scales, 196
Test taking, 14
Three Mile Island, 100
Total Quality Management (TQM), 79
Transistor, invention of, 37
Trigonometry, 284
   laws of cosines and sines, 287
   circular sine and cosine functions, 289

U

Units, 195
   derived, 195
   fundamental, 195
   systems of, 195

V

Veblen, Thorstein, 40
Venn diagram, 128
Visual aids, 126
Vita. *See* Résumé

W

Waterwheel, 32
Watt, James, 35
Whistle blowing, 96
Word processing, 257
Written communication, 117